THE BATTLE FOR GENESIS 1 AND 2

Creationism vs. Theistic Evolution

by

C. Wesley King

SCHMUL PUBLISHING COMPANY
NICHOLASVILLE, KENTUCKY

All scripture quotations, unless otherwise indicated are taken from the NIV. Permission to quote from the following scripture version is acknowledged with appreciation: *The Holy Bible, New International Version* (NIV), copyright © 1973, 1978, 1984 by the International Bible Society. Used by permission. All rights reserved.
Other versions used:
NKJV— New King James Version, copyright © 1982 by Thomas Nelson Inc. Used by permission.
ESV— English Standard Version, copyright © 2001 by Crossway. Used by permission.
NAS— New American Standard, copyright © 1978. Moody Press. Used by permission.
NLT— New Living Translation, copyright © 1988, 1989, 1990, 1991, 1993, 1996, by Tyndale House Publishers, Inc. Used by permission.
KJV— King James Version (Authorized King James Version).
RSV— Revised Standard Version, copyright © 1964, by Harper and Row Publishers, Inc. Used by permission.
Map on page 42 adapted from National Geographic Magazine, vol. 223, No. 1, January 2013, by permission.

Published by Schmul Publishing Co.
PO Box 776
Nicholasville, KY 40340

Printed in the United States of America

ISBN 10: 0-88019-583-5
ISBN 13: 978-0-88019-583-6

Visit us on the Internet at www.wesleyanbooks.com, or order direct from the publisher by calling 800-772-6657, or by writing to the above address.

Contents

Acknowledgments

To evangelical scholars and scientists who are quietly contending "for the faith that was once for all entrusted to the saints" by upholding the integrity and meaning of the Scriptures, despite the temptation to compromise with recent scientific assumptions concerning the origin of the world and man.

To Scott Huse for the use of the diagram the so-called Geologic Column and Timetable of Earth's history.

To John Hultink for his splendid essay entitled "Just Say the Word," which is gratefully used by permission in the last chapter.

Glossary

Anthropic principle— The idea that our universe is uniquely tuned to give rise to humans.

Archaeology— The scientific study of material remains (like fossil relics, artifacts and monuments) of past human life and activities, discovered by excavating predetermined sites or locations anywhere on earth. The remains of the culture of a people.

Astrophysics (1890)— A branch of astronomy dealing with the physical and chemical constitution of celestial matter.

Atomic physics— See Carbon 14.

Big Bang Theory (1955)— A theory in astronomy that the universe originated billions of years ago in an explosion from a single point of nearly infinite energy density.

Biology (1813)— A branch of knowledge that deals with living organisms and vital processes, whether plant, animal or human life.

Carbon 14 (1936)— A heavy radioactive isotope of carbon, mass number 14, used especially in trace studies and in dating archaeological and geological materials.

DNA (1953)— Serves as the information storehouse for a finely choreographed manufacturing process in which the right amino acids are linked together with the right bonds in the right sequence to produce the right kind of proteins that fold in the right way to build biological systems.

Epigenesis— The constant increase of the complexity of forms over a period of time. Something new is constantly added; something is additive to evolution, something quasi-creative.

Geology (1735)— A science that deals with the history of the earth and its formation.

Geologic time (ca. 1909)— The so-called long period of time occupied by the earth's geologic history. The "geologic column" or table is based on evolutionary presuppositions of long periods of time, namely the Precambrian Era, the Paleozoic Era (Age of Ancient

6

Life), Mesozoic Era (Age of Middle Life), and Cenozoic Era (Age of Recent Life), all calibrated in millions of years. (See also Uniformitarianism.) The "geologic column" is a fabrication. There is no proof for it.

Kosmos— The Greek for cosmos. In Homeric Greek it expressed fundamentally the thought of order, congruity and harmony in the make-up or arrangement of something. Eventually, it came to signify the created world, material universe, especially the earth; humanity, the people of the earth; human society, that ordered complex of human institutions, cultures and practices; and ethically, morally, or spiritually, the present world order or system of life in its condition of self-reliance and independence from God along with its apostate beliefs, values, ethics, attitudes, and practices. The *kosmos* as such is concerned with the present, transitory dimension of life, ignoring or denying eternity, gauging existence by material standards and defining achievement in terms of human esteem. It is a philosophy or worldview which in its crasser forms is marked by selfishness, greed, power, lust, ambition, violence, self-glory, and other vices which reflect the man-centered spirit of this world. In Colossians 2:20, to live in the world means to live as unredeemed captives of the fallen world who have not escaped its grip.

Legend— That which may have taken place regarding a person or thing.

Materialism— A term meaning that Matter (i.e., the fundamental particles that make-up both matter and energy) is all there is in the universe. A belief that everything observed in nature must have a natural explanation. This rules out miracles.

Materialist— To a materialist, reason begins by *assuming* that "in the beginning were the particles," and that the mind of man itself is simply a product of matter alone.

Myth— A usually traditional story of ostensive historical events that serves to unfold part of the worldview of a people or explain a practice, belief, or natural phenomenon.

Naturalism— The belief that nature is all there is in the world.

Newspeak— In his book *1984*, George Orwell introduced a concept called "newspeak," in which persons in positions of power began using terms and phrases that sounded right to the masses— when in fact they meant something very, very different. The same thing is taking place in some Christian circles today with regard to biblical and theological terms, like "the authority of God's Word" and "the inerrancy of the Scriptures."

Occam's Razor (William of Occam, 1836)— A scientific and philosophic rule that entities should not be multiplied unnecessarily, which is interpreted as requiring that the simplest of competing theories be preferred to the more complex, or that explanations of unknown phenomena be sought first in terms of known quantities. This rule may be applied to Creation vs. Evolution.

Progressive Creationism— The concept that the six days of Creation correspond to the geological ages, during which God was "creating" all things. According to this theory, there are certain admitted gaps in the fossil record, and these correspond to acts of creation; at other times, the created kinds were developing into their various families and genera. Also known as the Day-Age Theory.

Reductionism— is the effort to explain the complex by the simple, and the higher by the lower.

Science— Knowledge covering general truths or the operation of general laws, especially as obtained and tested through the scientific method. Such knowledge is concerned with the physical world and its phenomena. A branch of study which is concerned either with a connected body of demonstrated truths or with observed facts systematically classified and more or less colligated by being brought under general laws, and which includes trustworthy methods for the discovery of new truth within its own domain.

Scientific method (1854)— Principles and procedures for the systemic pursuit of knowledge involving the recognition and formulation of a problem, the collection of data through observation and experiment, and the formulation and testing of hypotheses.

Scientism (1877)— *(Broader meaning)* Methods and attitudes typical of or attributed to the natural scientist. An exaggerated trust in the efficacy of the methods of natural science applied to all the areas of investigation as in philosophy, the social sciences, and the humanities. *(Narrower meaning)* The belief that modern naturalistic science (evolutionary theory) is the great unifying answer to the most basic questions of human life, which would include questions of origin, purpose and destiny of human life. (See Humanistic Secularism.)

Sci-fi (1955)— Of, or relating to, or being science fiction. A sci-fi story of something that may or may not happen in the future. In terms of biology, reputable biologists believe that eventually they may be able to substitute normal genes for defective ones and thus cure sickle-cell anemia, hemophilia, and other types of hereditary disease.

Theistic Evolution— affirms that the biblical God was the Creator of all earthly organisms, humanity included, and used as his method the standard evolutionary scenario of gradual natural selection among genetic mutations across eons, according to geneticist Francis Collins and others.

Threshold Evolution— asserts that there is a wide possibility of change within the kinds originally created by God, but these variations cannot step over prescribed boundaries.

Uniformitarianism— The concept that the present is the key to the past. Processes now operating to modify the earth's surface are believed to have been operating similarly in the geologic past; that there is a uniformity of processes past and present. This theory denies the possibility that the earth's crust has been formed rapidly in a relatively short period of time during a worldwide disaster like the Genesis Flood.

Worldview— It is the belief and understanding that a person has of the world around him, or reality and life. There are only two worldviews: the *secular*, which is man-centered and based on naturalism and atheism, and the *biblical*, which is God-centered and based on the teachings and authority of the Scriptures. Conflict arises when one attempts to mix elements of one worldview with those of the other. We must be consistent in our belief system.

Foreword

ENESIS 1 AND 2 provides the believer with an account of God creating His universe *ex nihilo* ("out of nothing"). He spoke and all that exists came into existence. He created man from the dust of the earth, and man became a living being.

Now, when it comes to the issue of origins, we have two ways of understanding or explaining the universe and all that exists. It was either as the Bible declares or as some teach— a result of strictly naturalistic processes in evolution.

I find most students have a very unclear picture of evolution. I will hear it defined as "change over time." Creationists believe that! Variation around a mean is an observable fact. Natural selection was first defined by creationist Edward Blyth. In the scientific sense it means something quite profoundly different. Evolution is the process whereby inanimate matter became living matter by random chance through the mechanisms of natural selection and mutations, so that one living cell came into being that became the common ancestor of all that is.

This of course is profoundly different than the special creation by Divine fiat described in Genesis 1 and 2, and corroborated throughout sacred scripture.

So what are you going to believe? In today's mental environment students are taught evolution is a fact. We are told by many that science has disproved the Bible. Could you stand up to such a challenge as a believer? Our kids are being asked to do so.

For example, our students are told that the fossil record proves evolution is true. Is that accurate? Could you challenge that? Darwin himself said that the proof of evolution would come in the fossil record when "intermediate" fossils would show one species evolving into another. Do you know Darwin was concerned *not* to be able to identify such intermediate stages and since his research for the *Origin of the Species* that has not changed? Harvard paleontologist Dr. Stephen Jay Gould called it "paleontology's little trade secret."

Or consider the evidence of the human genome discussed in this book. Is this proof of evolution? Does it demonstrate man did descend from apes because the human/chimp genome is similar? Does it prove we have a common ancestor? Or does it prove we have a common Creator?

Are the results of science so assured that we cannot challenge them? Hasn't the history of science demonstrated the knowledge produced by scientific inquiry changes dramatically over time? Christian students absolutely need to challenge science. It is our right and responsibility when it attempts to discredit God's Word!

I find it interesting that it is not just Biblical creationists who find evolution unconvincing. Secular scientists in growing numbers find evolution a theory in crisis.

Yet it persists. Why? Because scientists *want* it to be true. It is their presupposition as much as the Biblical record is mine. Their careers and reputations are at stake. There is no room for a Creator in their thinking or commitments. Evolution as a proposal for the explanation of life did not originate with Charles Darwin. It is an idea as old as humankind to explain life without reference to a Creator.

The strangest position of all in the origins debate is Theistic Evolution, which Dr. King discusses in this book and which has made deep inroads into Christian colleges, universities and seminaries. This accommodation of the Biblical text to the scientific mindset is a theological disaster. The problem with the Biblical account of origins is not that it is esoteric or mythological or allegorical. The problem with the Bible account is it is clear.

For Theistic Evolution to replace special creation requires Theistic Evolution to revise the Scripture to fit the meaning of evolution and billions of years needed for evolution to be true into the Bible. The accommodation does not work.

For example, for Theistic Evolution to work within the Bible you have to make the following changes: the straightforward reading of the six days of creation does not work, the Biblical chronology does not work, and Adam and Eve cannot be the sole progenitors of the human race, the Flood is not true, nor the Tower of Babel. Are you willing to jettison all of that from the Bible? If the foundation of the

Bible in Genesis 1 and 2 is not true, the whole of the Bible becomes suspect.

Please read Dr. Wes King's book, *The Battle For Genesis 1 and 2,* carefully. It is a good book and will help you understand the battle and how to respond to the challenges. As he says, "The stakes are high." The church must grow in its understanding of just how important this debate is.

Thank you, Dr. King, for standing on the Word of God and for your faithfulness in helping a generation understand the complexities of this issue.

—GREGORY V. HALL, ED.D.
President, Warner University
Lake Wales, FL

Introduction

E VER SINCE THE RISE of higher critical scholarship in the 1700s and the appearance of Darwinian Evolution in the 1800s, and their attack on the veracity, historicity and authority of the Bible, particularly the Old Testament, *Genesis 1 and 2,* which deals with divinely-revealed truth about the origin of our universe and man, has borne the brunt of this attack because it does not conform to so-called scientific data and assumptions about beginnings. Even some who are willing to accept this divine truth with reservations have subjected the truth recorded in these two foundational chapters to theoretical adulterations and interpretations based on evolutionary presuppositions.

When evolution broke on the world scene in 1859 it was welcomed by the unbelieving world, but the Christian world (Protestantism) became sharply divided. As Howard Vos observes,

> The reaction of established religion [Christendom] was threefold: some, namely liberals, capitulated 'to evolution' and turned their backs on Christianity; others, fundamentalists for the most part, repudiated the new claims of science. The majority, Bible-believing Christians, worked out some sort of compromise between their faith and the new science.[1]

This division among Christians over origins and the role of evolution in this matter has remained to this day. On the one hand, I believe the fundamentalists were right in taking a stand against evolution. On the other hand, those who favored a mediating position between what Genesis 1 and 2 affirms about creation and what the new science was theorizing about, opened themselves up to unbiblical theories such as progressive creationism and theistic evolution. The net result in many cases was doubt and disbelief about revealed truth in Genesis 1-11 in general and Genesis 1-2 in particular. In some cases, this compromise or attempt to harmonize science and the Scripture has led to outright defection and departure from revealed truth.

The liberal and conservative reactions to evolution in the nineteenth century led in part to the Modernist/Fundamentalist controversies in the early twentieth century (1900-1940). Harry Emerson Fosdick was one of the leading exponents of modernism, and J. Gresham Machen of the fundamentalist cause. The controversies were over the nature of the Scriptures, the nature of man, the nature of the world, and the nature of education. The fundamentalists believed not only in the plenary inspiration and inerrancy of the Bible, but also in a series of evangelical doctrines such as the substitutionary death of Christ on the Cross, the reality of eternal punishment, and the need for personal conversion.

Sensing the great need to get beyond the modernist/fundamentalist conflicts of the early twentieth century, some prominent evangelical churchmen and theologians formed in 1942 the National Association of Evangelicals (NAE)— a remarkable Christian Fellowship embracing representatives of a wide variety of Christian organizations and denominations. The purpose of the new entity was to give the newly emerging evangelicalism one conservative and united evangelical voice in matters such as missions, education, evangelism, social action and world relief for the nations of the Third World. The Statement of Faith contained the essential doctrines of the Christian faith, but did not have an article on the growing debate (subject) of creation/evolution, namely that the universe and man were the creation of God alone. During the next four decades (1942-1980), the new evangelicalism grew in strength and prominence due to the support and influence of such leading advocates as Billy Graham, Bill Bright and Jerry Falwell, and the founding of new evangelical organizations such as the American Center for Law and Justice (ACLJ) to counter the American Civil Liberties Union (ACLU) and universities (Oral Roberts and Liberty), for instance. In 1956 Billy Graham founded *Christianity Today*, the voice and flagship of the newly emerging evangelical cause in America. Carl F. H. Henry was its first editor.

Shortly after the founding of the NAE in 1942, the American Scientific Affiliation (ASA) was created in 1944 to bring together Christian scientists who were committed to the belief that the Bible is the inspired Word of God and who, through their meetings and journal, could discuss many thoughtful proposals of a pathway toward harmony between science and faith. But within fifteen years of its formation, the ASA had developed a spectrum of belief in evolution and had become a strong proponent of Theistic Evolution. This is the belief that God acted as Creator and that He used evolution, with all that that implies, as His method of creating the material universe and man.

In 1966 Dr. Henry Morris wrote this assessment of the status of the creation/evolution debate at that time:

> Theistic evolution has, of course, been generally adopted in modernistic and liberal churches and seminaries for almost as long as Darwinism has been popular among scientists. Fundamentalists and other conservative schools and churches have, for the most part, reacted healthily against these trends and have maintained a vigorous insistence on the full reliability of the Biblical account of origins by special creation.

> But especially since the termination of World War II, with the rise of new-evangelicalism and the desire of erstwhile fundamentalists to attain intellectual recognition from the world, no doubt with the sincere desire to win more of the educated classes to conservative Christianity, there has come a continually increasing accommodation to theistic evolution in the thinking of these people.

Providentially, the 1950s, '60s and '70s witnessed a resurgence of Biblical Creationism with the publication of a flurry of books on Genesis, Creation and the Flood, and the founding of creationist organizations like the Creation Research Society in 1963 and the Institute for Creation Research in 1972. Despite these admirable efforts to reestablish orthodox beliefs concerning the special creation of man in the public mind, Theistic Evolution as a "mediating position" between evolution and creationism continued to be popular among Americans in general and attractive to Christian intellectuals in particular. Polls in the 1990s revealed that ninety percent of Americans were evenly divided between Biblical Creationists and Theistic Evolutionists.

There can be little doubt that the American Scientific Affiliation (1944) has had a growing impact on the thinking of some science professors (principally biology) and theologians in Christian colleges and seminaries over the past several decades. A recent non-random Internet survey of teachers at evangelical seminaries in 2009 showed that forty-six percent accept the concept that the biblical God was the creator of all earthly organisms, humanity included, and used as his method the standard evolutionary scenario of gradual natural selection among genetic mutations across eons. Karl Giberson estimates that "the overwhelming number in biology departments at Christian colleges would be fine with this," as well.

It is now obvious that there is a serious crisis in evangelical circles. This present crisis has been brought on by several recent developments: (1) the full mapping of the human genome in 2003; (2) the full mapping of the DNA and chimpanzee genome in 2005 by the National Institute of Health (NIH) headed up by Francis Collins; (3) the "near identity" of the human and chimp genomes; (4) the publication of Collins' book *The Language of God* in 2006, in which he is sharply critical of creationists and passionately defends his version of Theistic Evolution; (5) Collins' launching of the BioLogos Foundation in 2007 to promote Theistic Evolution, especially among evangelicals; and (6) the rush of Christian college professors and theologians to join his cause.

Collins, one of the most eminent scientists ever to identify as an evangelical Christian, staunchly defends Darwinian Evolution even as he insists on God as the Creator. He now stands at the epicenter of a dispute with fellow believers over the traditional biblical view that God directly created Adam and Eve, the historical parents of the entire human race.

The stakes in this controversy are extremely high. In previous decades, theologically liberal Christians, atheists and evolutionists have gone to great lengths to undermine the authority and trustworthiness of the Scriptures. But now, those who should be holding the line on the complete integrity and teaching of the Scripture, are now toying with, or even believing in, a different origin of man. This theory not only goes to the heart of the Christian belief of who man really is, but it goes to the heart of the Gospel itself. This is why I am calling this *The Battle for Genesis 1 and 2* and why I'm writing this book. I believe that Satan, the archdeceiver and enemy of God, man, and the Bible, is now going for "the jugular" of Christianity, because he sees his end approaching. The future of evangelicalism hangs in the balance. Which worldview are we going to cling to—some newfangled and unproven biological discovery, or the revealed truth in Genesis 1 and 2?

My Objectives in Writing About this Critical Issue

First, to *explain* to Christian lay people and youth what Theistic Evolution is.

Second, to *inform* the Christian community about this dangerous departure from the heaven-sent truth recorded in Genesis 1 and 2, which we wouldn't have known about had God not revealed this to Moses for us.

Third, to *encourage* Christians to be familiar with the arguments for Biblical Creationism versus the fallacies and deceptions of Theistic Evolution.

Fourth, to *strengthen* our faith in God, His person, power, promises and revealed Word— "The faith that was once for all entrusted to the saints *(hagios)*," Jude 3. In this passage we are urged to contend for this faith. Why? Because of false teachers and false teachings.

Fifth, to *pray* for Christian college presidents and administrators that they will be careful in hiring faculty who believe in Biblical Creationism without reservation, as recorded in Genesis 1 and 2.

If we are going to raise up college and seminary-trained ministers who will preach the Word with power, they must not only be filled with the Holy Spirit, but they must also believe wholeheartedly in the inspiration, inerrancy, and authority of God's Word and that God means what He says in that Word in understandable language.

Sixth, to *grow* an "army" of Creationists to disarm Theistic Evolution and its deceptive hermeneutic.

A Brief Overview of the Book

Obviously, we have not arrived at this crisis in evangelicalism overnight. This crisis has been developing incrementally for several centuries.

Part 1: Theistic Evolution— The Liberal view (From Charles Darwin to Francis Collins)

An overview of the seventeenth, eighteenth and nineteenth centuries. Chapter one: "The Undermining of Divine Revelation by the

Enlightment and Higher Criticism in the Seventeenth and Eighteenth Centuries."

Chapter two: "The Appearance of Evolution in the Nineteenth Century (1800-1900)."

Chapter three: "The Rise and Spread of Theistic Evolution from 1859 Onward."

Chapter four: "A Timeline of Growing Evangelical Acceptance of Theistic Evolution."

Chapter five: "A Biblical Response to and Critique of Francis Collins' book, *The Language of God*."

Chapter six: "A Biblical Response and Critique (continued)."

Part 2 is devoted to **The Role of Science in Relation to Revealed Truth about Creation.**

Chapter seven: "Science: Origin, Disciplines, Scientific Method and Departure from Divine Truth."

Chapter eight: "Science: Definition, Limitations and Agreement with Creation Events in Genesis 1 and 2."

Part 3: **Foundations for Interpreting Genesis 1 and 2**

Chapter nine: "Foundations for Interpreting Genesis 1 and 2."

Chapter ten: "The Interpretation of Genesis 1 and 2 Using the Inductive Method; What does it say and What does it mean?"

Chapter eleven: "Conclusions Based on an Analysis of Genesis 1 and 2."

Chapter twelve: "Wanted: A Centurion-like Faith: "Just Say the Word."

Afterword: "Four Illusions of Theistic Evolution."

Who is This Book For?

This book is written to inform thousands of ordinary laypersons and their grandchildren in our various denominations of the current crisis in evangelicalism. I firmly believe in the power of the gospel (preached Word) to transform human hearts and minds. I also believe in the ministry of the Holy Spirit in guiding new believers into all truth, not only about our redemption, but also about our creation. I am an optimist, believing the sovereign God stands in charge of our world, confused though it is, and knowing that when history comes to its endpoint, believers will cry, *"Christus victor!"*

But I also believe that this is a time for action. Let us close ranks and in a spirit of prayer and humility, be ready to witness to those who believe and teach another doctrine other than that God is the sole and sovereign Creator of all things.

—C. Wesley King, D. Min.

Lakeland, FL

Endnote

1. Howard Vos. Exploring Church History (Nashville: Thomas Nelson Publishers, 1994), pg. 120.

PART 1
THEISTIC EVOLUTION– THE LIBERAL VIEW
From Charles Darwin to Francis Collins

1

The Undermining of Divine Revelation by the Enlightenment and Higher Criticism

"Did God really say... ?"

THE PROBLEM OF DOUBT and disbelief of God's Word is not new. It is as old as human history. The problem began in the Garden of Eden. Soon after God had created the first man and the first woman in His moral and spiritual image, He gave them a command to test their love and loyalty for their Maker. The command was very simple and clear, and contained ample freedom with just one prohibition. "You are free to eat from any tree in the garden; but you must not eat from the tree of the knowledge of good and evil, for when you eat from it you will surely die" (Gen. 2:17). This short command from God may be considered the germinal seed of the divine revelation of God for the salvation of humanity now contained in the Bible from Genesis to Revelation. Of a truth it was the first Word of God that was being communicated to the first human couple. What is most significant about this command, in the second place, is that the serpent, who was later identified by the Apostle John in Rev. 12:9 and 20:2, 7, as the devil or Satan, also knew about the command.

We know this to be true because Genesis chapter three opens with these words: "Did God really say, 'You must not eat from any tree in the garden'?" (Gen. 3:1). Several things stand out here. *First*, the serpent's question was designed to cast *doubt* upon what God had said. *Second*, the question was phrased in the form of a half-truth because the enemy left out the prohibition, "but you must not eat from the tree of the knowledge of good and evil, for when you eat of it you will surely die." The dictionary defines a half-truth as a statement that mingles truth and falsehood with deliberate intent to deceive.[1]

One commentator captured the essence of the serpent's deceptive move this way: He [the tempter] followed this with a flat denial [lie] of the divine warning, "You shall not surely die." While Eve's response to the serpent gave little indication of her taking the bait, he followed up his first subtle suggestion with a bold attack which left her little question as to his opinion of God. Then to the insidiousness of doubt, half-truth and the sacrilege of attempting to rob God of His truthfulness, he added the blasphemy of suggesting that God's command was due to ulterior and selfish motives, "for God doth know that in the day you eat thereof, then your eyes shall be opened and ye shall be as God, knowing good and evil."[2]

The serpent's intent was to get Eve to disobey God's command. The sacred writer declares, "When the woman saw that the fruit of the [forbidden] tree was [1] good for food and [2] pleasing to the eye, and [3] desirable for gaining wisdom, she took some and ate it. She also gave some to her husband, who was with her, and he ate it. Then the eyes of both of them were opened, and they realized they were naked; so they sewed fig leaves together and made coverings for themselves" (Gen. 3:6-7).

Doubt concerning God's Word gave way to disbelief, disobedience, broken fellowship with their Creator, disharmony in their lives, and disorder in society and the world.

Legitimate versus Illegitimate Doubt

There may be times in life when it is legitimate to doubt something. The story has been told that Christopher Columbus, for instance, doubted the prevailing theory that the world was flat, and insisted on testing his own belief in the roundness of the earth. Today, someone may be diagnosed with a certain medical condition, but they have their doubts about the diagnosis. So they go to another doctor and ask for a second opinion as to the correctness or authenticity of the first opinion. That is certainly understandable and a person's privilege.

But when it comes to the Word of God— the Bible— there is no place for doubt as to its absolute truthfulness and trustworthiness in every part. This is true because the same God who created the universe, earth and mankind "in the beginning," and was fully cognizant of man's fall into sin, also revealed the conditions in primeval history (Genesis 1-11). He inspired holy men of old, prophets in the Old Testament and apostles in the New Testament, to write down His plan of salvation for humankind progressively over 1600 years until the "fullness of time" (Gal. 4:4) when God sent His one and only Son into the world to be man's Redeemer. Since the coming of Jesus Christ and His return to the Father, He has become the central figure of history and Living Word (John 1:1-3; 14-16) within the written Word of God. The Word of God from Genesis 1:1 to Revelation 22:21 is a perfectly revealed and divinely executed plan of salvation— a harmonious unity of truth that cannot be broken (John 10:35).

The Age-long Conflict Between God and Satan

When Satan was expelled from heaven and fell from his lofty position as Lucifer, son of the morning, he became the "god of this age" (2 Cor. 4:4), the "prince of the power of the air" (Eph. 2:2), and the "prince of this world" system (John 14:30). The earth became the special sphere or territory of Satan's evil activities and influence.

Christians today must remember that the eternal God of the ages (of power, holiness and goodness), and Satan, the father of lies and archdeceiver of men and nations, are in mortal and spiritual conflict with each other. On the one hand, God's purpose is to establish His kingdom of righteousness on earth in the hearts of men and women; to transform persons, societies and nations by His renewing and sanctifying grace. On the other hand, "The worldwide and age-long works of Satan can be traced to one prevailing motive. He hates both God and man and does all within his power to defeat the plan of God in order to establish his own kingdom of evil [on earth]. His overshadowing motive seems to be that of gaining equality with God through the means of deception."[3] At the heart of this battle is the Bible— the living and enduring Word of God. "For all men are like grass, and all their glory is like the flowers of the field; the grass withers and the flowers fall, but the word of the Lord stands forever" (1 Pet. 1:23-25). The strategy of Satan from the Garden of Eden to the present has been to entice men and women in general, and people of faith in particular, to *doubt* and *disbelieve* the Word of God.

The Modern Assault on the Authority and Inerrancy of the Bible.

During the Pre-modern Era of church history and western thought (AD 1-1700), the key word was *revelation*. The foundational belief of men and scholars during this historical period was that God had revealed certain truths, called ultimate truths, that were inaccessible to human reason. These ultimate truths concerned the existence of God; the creation of man and the world by God along with a belief in a relatively young earth; the purpose of man on earth; the Virgin Birth, life, death and resurrection of Jesus Christ; the redemption of human beings from sin by the atonement of Christ on the cross; and the final destiny of humankind and the world. The foundation of these truths was the Bible— the written Word of God— and Jesus Christ, the Living Word of God. The center of this period was God.

The advent of the Enlightenment Age (1700-1800) turned all this on its head, by dethroning *revelation* and enthroning *human reason* in the Modern Era (1700-2000). *Rationalism* is the philosophy that exalts human reason (darkened by the Fall into sin) and man's inter-pretation of life— origins, nature, purpose and destiny— over the divinely-revealed explanation of it recorded in the Bible.

The modern assault on the authority and integrity of the Scriptures is known broadly as "biblical criticism." This criticism is twofold: Higher Criticism and Lower Criticism. Higher Criticism, largely destructive and liberal, seeks to destroy the supernatural nature of the Bible (its miracles and inspiration by the Holy Spirit), whereas Lower Criti-cism, which is constructive and conservative, is the study of the text of the Bible in an attempt to ascertain whether the text that we have is the one which came from the hands of the writers.[4] It has been the radical Higher Criticism rather than the Lower Criticism that has de-stroyed many persons' faith in the divine revelation in the Bible.

Higher Criticism of the Old Testament

The beginning of Higher Criticism is associated with the name of Jean Astruc (1684-1766), an eighteenth century French doctor, who in 1753 divided the book of Genesis into two parts. He *assumed* the use of two documents as sources because he found the name Elohim (God) in some places and Jehovah (Lord) in others.

Hermann Hupfeld (1769-1866) another OT critic in 1853 was the first to claim that the Pentateuch was the work of different authors rather than a single narrative composed from many sources by Moses. Karl Graf and Julius Wellhausen developed a well-elaborated *theory*, known as the Graf-Wellhausen theory, which has been adopted by the higher critics. According to this theory, the sections in which the name Jehovah (J) is used is the earliest document; another part by another author is E (Elohim); still another is Deuteronomy as D; and P as Priestly. In this fashion the unity of the Pentateuch and its Mosaic authorship was denied.

The Tubingen and Wellhausen schools of religious thought in Germany were two that subscribed to the evolutionary and higher critical viewpoint in religion. Julius Wellhausen (1841-1918), a key figure in the rise of liberal scholarship, denied Mosaic authorship of the Pentateuch, concluding that it was post-exilic (after 516 BC) and believed that the Old Testament was put together by later editors using a variety of source materials. During this time, it was commonly taught that man started out with no religion and finally advanced to the elevated viewpoint of monotheism a few centuries before Christ. As Howard Vos observes, "He [Wellhausen] applied to religion and the OT the same evolutionary principles that Darwin and others were applying to the natural sciences including biology. The system he constructed was destined to have impact worldwide during the twentieth century."[5]

Later critics divided Isaiah into at least two parts, 1-39 and 40-66, thus robbing the greatest OT prophet of his prophetic insight with regard to Israel's future, which he had received by the inspiration of the Holy Spirit. Higher critics also lowered the date of Daniel to the Maccabean period (166-40 BC) so that it became history rather than prophecy and history as fulfilled prophecy.

Higher Criticism of the New Testament

The beginning of Higher Criticism of the New Testament is usually associated with the name of Hermann Reimarus (1694-1748), who taught Oriental languages at Hamburg. In his *Fragments* (1778) he denied the possibility of Biblical miracles and advanced the idea that the writers of the NT with their stories or miracles were pious frauds. Elaborating on this, Gotthold Lessing argued that the Scriptures served man as a guide during the primitive phase of his religious development, but that *reason* and duty were sufficient guides in the more advanced state of religion.[6]

> With regard to the New Testament, Earle Cairns observes that Ferdinand Baur (1792-1860), who borrowed Hegel's logic [thesis, antithesis and synthesis] argued in 1831 that in the early Church [AD 33-100] there had been a Judaism that emphasized the Law and the Messiah. This earlier approach can be observed in the writings of James. Paul developed an antithesis in such books as Romans and Galatians in which the emphasis was upon grace rather than upon law. The old Catholic Church of the second century represented a synthesis of Petrine and Pauline views. This synthesis is revealed in such books as the gospel of Luke and the Pastoral Epistles. Baur then proceeded to date the books of the New Testament in this Hegelian framework as either early or late according to the manner in which they reflected Petrine, Pauline, or synthesizing influences. Historical data gave way to philosophical presuppositions in ascertaining the chronology of the books of the New Testament.[7]

Some theologians, who adopt critical views of the New Testament, consider that the essence of the Gospel is in the ethical teachings of Jesus and that Paul changed the simple, ethical religion of Jesus into a redemptive religion. Destructive Higher Criticism has led many to deny the inspiration of the Bible as a revelation from God through men inspired by the Holy Spirit, and to minimize or deny the deity of Christ and His saving work upon the Cross of Calvary. *The Life of Jesus* (1835-1836) by David Strauss (1808-1874) combined all of these views. Strauss denied both the miracles and integrity of the New Testament as well as the deity of Christ.

Sadly, Cairns observes,

> Germany, once the home of the Reformation, became the land in which criticism developed. The history of modern Germany well illustrates the lengths to which men will go when they deny God's revelation in the Bible and when they replace *revelation* with *reason* and *science* as the authority for thought and action.[8]

Sadder still is the fact that these liberal and higher critical views of the Bible, Christ and the Christian faith were introduced into the universities and seminaries of the northeast and other institutions across America in the latter nineteenth and twentieth centuries. This defection from the biblical and orthodox understanding of God's revelation in the Bible has influenced myriads of pastors and churches, and led to the rise and fall of higher education in America. (See the introduction to the book *Already Compromised* by Ken Ham and Greg Hall.)

Summary

In this chapter we pointed out that Satan's strategy has always been to get men and women and people of faith to *doubt* and even to *disbelieve* what God has said in His Word— the Bible.

During the Pre-modern Era (AD 1-1700) of Christianity, the Church accepted the body of ultimate truths revealed to prophets and apostles as contained in the Holy Scriptures.

The Enlightenment Age (1700-1800) changed this orthodox situation by dethroning *revelation* and enthroning *reason* and *science* in the Modern Era (1700-2000). As noted earlier, rationalism is the philosophy that exalts human reason and man's interpretation of life over the divinely-revealed explanation recorded in the Bible. In contrast to Pre-modern beliefs, a modern understanding of the world leaves no room for God and the supernatural. Reason and the scientific method, developed by Francis Bacon, take over as the dominant interpretation of life. In modernity, the foundations of truth are science and rationalism.

The assault on the authority and integrity of the Bible began in the Enlightenment period. This modern assault consisted of higher critical attacks on the Bible as a whole, its miracles, message and divine inspiration.

In the nineteenth century, higher critics like Julius Wellhausen undermined the integrity of the Pentateuch, and Genesis in particular, by denying its unity, Mosaic authorship, and by considering the stories in Genesis mere myths and legends.

Commenting on the above developments in Bible scholarship, Cairns asserts,

Criticism of the Bible, Darwin's theory of evolution [which will be discussed in chapter two] and other social and intellectual forces created religious liberalism in the late nineteenth century. Liberal theology has been greatly devoted to the scientific method and has applied evolution to religion as a key that might explain its development. It has insisted upon the continuity of man's religious experience to such an extent that the Christian religion has become the mere product of a religious evolution rather than a revelation from God through the Bible and Christ. Christian experience has been emphasized much more than theology. These views of liberalism conservative Christianity has fought, and the movement associated with the name of Karl Barth has opposed.[9]

It was the struggle between theological liberalism and conservative fundamentalism in the early twentieth century that gave rise to the neo-orthodox movement, championed by Karl Barth and Emil Brunner, as a mediating position between these two warring factions. The strengths of neo-orthodoxy were: 1) the renewed emphasis on the sinfulness of man and his need for divine grace, 2) the centrality of Christ, and 3) the importance of justification by faith once again. Its weaknesses were consequential: 1) their dropping of the formal principle of the Reformation, namely, the inspiration of the Scriptures, 2) their denial of propositional truth in the Bible, 3) their rejection of objective truth in the Bible, and 4) their insistence that the Bible only becomes the Word of God to the reader in the divine-human encounter.

At the same time, Rudolph Bultmann was proposing religious existentialism and the need to "demythologize" the New Testament so that the modern reader could better understand its teachings.

This perplexing and confused theological environment in the mid-twentieth century called for a true evangelical and theological expression of Christianity based on the timeless truths of the Word of God. Thus the founding of the National Association of Evangelicals in 1942 and the broad evangelical movement of the second half of the twentieth century.

Now, after sixty years of marked spiritual influence on the nation and the world, and remarkable achievements, it seems that neo-evangelicalism is losing its way and compromising fundamental biblical beliefs that it has long held. Some evangelical scholars in Christian institutions and scientists are now undermining the timeless and revealed truths of Genesis 1 and 2 by calling into question the divine origin of man in their search for the historical Adam and by giving credence to the so-called near identify between the recently encoded chimp genome and the human genome.[10] For this reason we have given this book the title of *The Battle for Genesis 1 and 2.*

As Old Testament scholar Eugene Carpenter has affirmed so correctly,

> For if he [man] is not who and what the Scripture declares he is, the teleology [the ultimate purpose] of the cosmos [universe and world for which God so loved and gave His one and only Son to redeem] is shown to be vacuous [empty of meaning].[11]

The very physical, moral and spiritual nature of man is now at stake. More will be said about this in later chapters.

Endnotes

1. *Webster's Ninth New Collegiate Dictionary*, (Springfield, MA: Merriam Webster Inc. Publishers, 1986), 548.

2. Charles Carter, ed., *The Wesleyan Bible Commentary*, I (Grand Rapids: William B. Eerdmans Publishing Company, 1967), 35.

3. Charles Carter, ed., *A Contemporary Wesleyan Theology*, 2 (Nicholasville, KY: Schmul Publishing Co., 1992), 1083.

4. Earle Cairns, *Christianity Through the Centuries*, (Grand Rapids: Zondervan Publishing house, 1958), 447.

5. Howard Vos, *Exploring Church History* (Nashville: Thomas Nelson Publishers, 1994). 120-121

6. Cairns, 448.

7. Loc. Cit.

8. Ibid., 449.

9. Ibid., 452-453.

10. Richard Ostling, "The Search for the Historical Adam" in *Christianity Today* (June 2011), 23-27.

11. Eugene Carpenter, "Cosmology," in *A Contemporary Wesleyan Theology* (Salem, OH: Schmul Publishing Company, Inc. 1992), 178.

2

The Appearance of Evolution in the Nineteenth Century (1800-1900)

Introduction

IN CHAPTER ONE WE SHOWED how Higher Criticism of the seventeenth and eighteenth centuries went a long way in undermining the authority and integrity of the Bible. The nineteenth century (1800-1900) has been called the Age of Science. Science did not take over, though, until about the middle of the century. The Enlightenment of the eighteenth century had gone too far in its rationalism and in its efforts to discredit the Word of God and to eradicate religion and feeling from all of life. The first part of the nineteenth century witnessed the rise of Romanticism as a reaction to that extreme.

Romanticism was characterized by a new emphasis on feeling and faith— but not orthodox faith— individualism and communion with nature, divine and untamed. Ralph Waldo Emerson (1830-1882), disillusioned with his ministry in the Unitarian Church, left that church and became one of America's most noted and influential literary philosophers and romantic poets. For his views on God, nature and pantheism, see my book, *Creation for Earnest Believers*, ppg. 45-46.

At the same time, there was also a new emphasis on the organic view of history and society, namely, that there is slow, not radical, development of the social organism. As Howard Vos accurately points out, "This intellectual context is important for the appearance and impact of Darwinian thought."[1]

Predecessors to Charles Darwin

The new emphasis on reason as opposed to revelation found different expressions in Europe and America. In France, led by Rousseau and Voltaire, it took the form of *atheism*, the rejection of Christianity and the disastrous results of the French Revolution. In Germany, it was known as the *Enlightenment*. Christian Wolf, Hermann Reimarus and Gotthold Lessing were among its leading exponents. In North America, the most famous advocate was Thomas Paine, deist, who wrote *The Age of Reason*.

In England, rationalism took the form of *Deism*. Deism is sometimes called the "Watchmaker" theory, in that God was considered to be the necessary first cause to start the system going, but once He set the universe in motion, He no longer interfered with its natural processes. This amounted to an extreme view of God's transcendence to the exclusion of His immanence, or nearness, to His created beings and universe. The crowning feature of the English deists was the naturalization of religion whereby they eliminated from it all supernatural elements. Rationalism had penetrated the ranks of orthodoxy. Christianity was seen as a simple set of good ideas, having a form, but devoid of power to transform lives.

Another development in England at this time was the organization of the Lunar Society of Birmingham. This aristocratic group of fourteen men sought social change and the advancement of a secular society. R.E. Schofield observes that these members were some of the most influential men in England, and that their primary intention was to remove the Church from a position of power in Great Britain.[2]

The Lunar Society recognized that the Bible, as God's self-revelation to humanity, was the greatest single obstacle to the achievement of its socialistic aims. It concluded that generating *disbelief* in the Bible would be the most effective way of changing public opinion. But rather than casting *doubt* on such cherished doctrines as the Virgin Birth of Jesus Christ or His resurrection, the Lunar Society chose to discredit the biblical accounts of Creation and the Flood.[3]

The founder of the Lunar Society was a man named Erasmus Darwin (1731-1802), grandfather of Charles Darwin.

Erasmus Darwin's contribution to the emerging view of evolution was a two-volume work written in 1794-1796 called the *Zoonomia*. It expressed the essence of the theory that his grandson announced to the world five decades later.[4]

New Ideas about the Earth's Age

At this point I need to remind the reader that before the Age of Enlightenment/Reason around 1700, most Jews and Christians believed in the Genesis account of creation and a relatively young earth based on Archbishop James Ussher's date of 4004 BC for creation. Bishop Ussher (1581-1656) was a British scholar of vast learning and an authority on a wide range of subjects including biblical chronology.[5]

However, two English geologists, James Hutton (1726-1797) and Charles Lyell (1797-1875), challenged the biblical view of Creation and the Genesis Flood. Raised a Quaker, Hutton eventually rejected the belief in a literal worldwide flood. He argued that the earth's history could best be explained by examining the earth's layers rather than accepting the validity of questionable Jewish records.[6]

Hutton proposed that the earth had been molded, not by sudden violent events like the Genesis Flood, but by slow and gradual processes, the same processes that can be observed in the world today. This theory became known as *Uniformitarianism.* It implied that the earth has a long living history. Six thousand years is not enough time for such major evolutionary changes to take place. Caryl Matrisciana and Roger Oakland believe that Hutton's was the first prominent scientific voice that not only promoted a humanist view of the origin and history of earth, but boldly undermined the authenticity of the Bible. He expanded the accepted biblical time frame for the age of the earth and refuted the biblical account of a global flood.[7]

Charles Lyell, who is remembered for his significant contributions to the development of Uniformitarian geology, was born the year James Hutton died. He was a lawyer, politician, geologist and the author of *The Principles of Geology* (1830-1833). In this work he continues to cast doubt on the truthfulness of the Bible and to reveal his keen interest in evolutionary geology, meaning the earth evolved over long periods of time, even millions of years.

The seeds of doubt that had been sown by the Rationalists of the seventeenth century and were nurtured by the skeptics and Deists of the eighteenth century came to full flower in the nineteenth century in the persons of Charles Lyell and particularly Charles Darwin (1809-1882). In *The Ascent of Man*, Colin Brown asserts, "The whole fabric of Christianity was called into question. Science, philosophy and history were all called upon to show that the Christian faith no longer had a leg to stand on."[8]

The Appearance of Evolution as a Scientific Formulation

Evolutionary views of the origin of man and the universe date back to Aristotle and the Greeks more than 2500 years ago. But it was Charles Darwin who first attempted a scientific development of the philosophical theory of evolution with the publication of his book *Origin of the Species* in 1859. By this time the modern scientific era was coming into its own.

Towering above all others, Charles Darwin stands out in most peoples' minds as the father of evolution. Indeed, the terms *Darwinism* and *evolution* have become almost synonymous. So, a brief review of his life, theory and influence is in order at this point.

His Life— It is difficult to know how much influence, if any, Erasmus Darwin had on his grandson's thinking. What we do know is that at his father Robert's suggestion, young Charles studied medicine at Edinburgh for two years before moving to Christ's College in Cambridge to study theology and prepare for a life in the ministry. Darwin graduated from seminary in 1831. At the age of twenty-one he was about to take a small Anglican country parish when he was invited to join the *HMS Beagle* in its voyage around the world (1831-1836) on a survey expedition to explore the wonders of natural science. This is where his interests really lay. Instead of a pastor, he became a naturalist.

The course of his life was changed forever. This momentous voyage, which began in December of 1831, transformed Darwin's thinking. He describes his original point of view thus: "I did not then in the least doubt the strict and literal truth of every word in the Bible.[9] Whilst on board the Beagle I was quite orthodox, and I remember being heartily laughed at by several of the officers for quoting the Bible as an unanswerable authority on some point of morality."[10] Two years following the return of the *Beagle* to England, Darwin wrote the following:

> I was led to think much about religion. But I had gradually come by this time to see that the Old Testament from its manifestly false history of the world, with the Tower of Babel, the rainbow as a sign, etc. and from attributing to God the feelings of a revengeful tyrant, was no more to be trusted than the sacred book of the Hindoos, or the beliefs of any barbarian.[11]

In other words Darwin had come to despise and reject the Old Testament. What is the lesson for today's Christians? Just this: Never give up your faith and confidence in Genesis as divinely-revealed historical fact!

How were Darwin's religious views so drastically changed? Some claim it was the geological and biological observations he made while traveling around the world that caused him to doubt the authenticity of the Bible.[12] Others suggest that Lyell's *Principles of Geology*, which he read during the voyage, exerted a powerful influence on his thinking as the journey progressed.[13] Herman Hausheer believes that Darwin "lost his religion when he assumed that religion depended upon a definitive scientific view."[14] Whatever the case, he rapidly moved away from a belief in fundamental Christianity (theism) to agnosticism. By the late 1830s he had completely abandoned his original Christian faith.

His Theory— Darwin's voyage convinced him that differences between living animals and fossils on the main lands and those on the islands (principally the Galapagos, west of Ecuador) that he visited, could only be accounted for by biological evolution. In his two works, Charles Darwin offered solutions to two burning questions of his day. What precisely are the causal origins of living things? And, where did human beings come from? Having rejected the Genesis account of the origins of the universe and human beings, Darwin turned to fallen human reason to explain how life had developed. He published his first book *Origin of the Species* in 1859 after finding that Alfred Wallace had independently arrived at similar conclusions.

His general theory is that organic forms are the result of a long process of development from the most insignificant beginnings under the continued influence of the environment. Herman Hausheer writes:

> He opposed the theological and romantic view of men as fallen angels. He held the realistic view that man developed from an animal into a spiritual and moral being. Neither psychologically nor physically did he allow any but quantitative differences between man and beast. Darwin's scientific materialism is characterized by its mechanical explanation of the world, its absolute negation of final causes, and its denial of design. According to Darwin, man is the descendant of a favored variety of apes. [But] according to Genesis, our species sprang from a clod of earth, a much more humble origin than the origin from apes.[15]

Modern evolutionary theories are modified and updated versions of Darwinian thinking, hence the term Neo-Darwinism. Scott Huse accurately describes this newest point of view like this:

> Modern scientists, who propose this type of viewpoint, postulate that the combined effects of natural selection, mutations, and geologic time could account for organic evolution. Neo-Darwinists believe that mutations supply the needed variants which nature can preferentially select over eons of time. They concede that neither mutations nor natural selection alone could account for the supposed evolutionary progression of Life.[16]

Earle Cairns, long time history professor at Wheaton College, offers this fine analysis and rebuttal of Darwinism and Neo-Darwinism.

> In his book Darwin argued that the struggle for existence kept the population of the various species constant in spite of the fact that reproduction is geometric and that many more are produced than are essential for the survival of the species. In this struggle some individuals develop characteristics favorable to survival through a process of adjustment and adaptation to environment. These characteristics are passed on by sexual selection in which the favored males and females mate. Thus only the fittest survive. He thought that such a similarity as that of the body structure between man and animals substantiated this theory, but he forgot that this and other similarities might be evidence of *design* on the part of the Creator who gave His creatures similar body structures because of the similarity of their environment. Darwin applied his theory to man in *The Descent of Man* (1871) and argued that man was linked with animal life by common ancestral types.[17]

Cairns goes on to point out,

> Darwin's idea of continuity between man and animal has been summarized as "descent with change." This view is opposed to the Biblical concept of special creation by God with fixity in the groups thus created... No missing link that would conclusively identify man with animals has been discovered; in fact, cross-breeding between many groups is impossible... God is said to have made the different groups reproduce "after his kind."[18]

His Influence— with these two writings, Charles Darwin gave modern science a hypothetical (but not factual) and revolutionary way of thinking about the universe and everything in it. Without question, the arrival of Darwin's books changed the world.

It was not long before Darwin's theory on the origin and development of life became well known. Soon intellectuals began to apply the concept of biological evolution to their own fields of endeavor: geologists and paleontologists to the natural sciences, psychologists and sociologists to cultural interpretation, communists and others to social theory (Karl Marx compared the struggle for survival among organisms to the struggle for power among social classes) and biblical critics like Julius Wellhausen and David Strauss to the Christian religion. For example, evolution was used to justify the idea of race superiority, whether by Hitler and the Aryan race, or by the white supremacists in America, because the idea seemed to fit in with Darwin's concept of the survival of the fittest. It has also been used to justify having no absolute foundation or norm for ethics. Good conduct is merely those actions deemed suitable by each generation for the proper conduct of society. The doctrine of evolution has also been used to glorify war as the survival of the fittest. All of these conclusions have been reached by the application of biological theory to other fields through an unwarranted use of the argument from analogy. Consequently, evolutionary thinking based on theory alone has come to penetrate every area of our lives and society, particularly our institutions of higher learning. Even though evolution is based on unproven assumptions, it is taught and believed as scientific fact.

Although the theory of evolution denied the direct creation of man by God, the greatest damage came from the application of the theory to the development of religion. God and the Bible were looked upon as the evolutionary products of man's religious consciousness, and the books of the Bible were dated accordingly. The biblical eschatology, in which perfection would only come in this world by the direct intervention of God through the return of Christ, was replaced by the evolutionary view of a world that was becoming increasingly improved by human efforts.

Because man was not guilty through original sin there was no need of Christ as Savior. Christ was viewed merely as a great teacher and sin was seen as merely the remnant of animal instinct in man.[19] So, in the nineteenth century the Christian faith found itself challenged from three directions: from *science* in the shape of the theory of evolution; from *philosophy* in the form of alternative worldviews such as pantheism, deism and agnosticism, intended to make belief in God obsolete; and from *history* in the guise of biblical criticism.

Reaction to Evolution

When evolution broke on the world scene, as Howard Vos observes, "The reaction of established religion was threefold: some capitulated and turned their backs on Christianity [namely, theological liberals]; others repudiated the claims of science [fundamentalists, for the most part]; the majority worked out some sort of compromise between their faith and the new science [Bible-believing Christians] in general."[20] The liberal and conservative reactions to evolution in the nineteenth century led to the Modernist/Fundamentalist controversies of the early twentieth century in America (1900-1930). Harry Emerson Fosdick was one of the foremost exponents of modernism, and J. Gresham Machen of the fundamentalist cause. The controversies were over the nature of the Scriptures, the nature of man, the nature of the world, the nature of redemption, and the nature of education.

The third reaction to evolution resulted in compromises with Darwinian thought in various forms within the Church from Darwin's time to the present. These theories include: 1) Uniformitarianism, 2) the Gap Theory, 3) the Day-Age Theory, 4) Theistic Evolution, and 5) the Big Bang Theory, all of which are evolutionary-based interpretations and unwarranted assumptions with regard to the revealed text of Genesis 1 and 2, that cannot be proved. These will be dealt with in greater detail in chapter eleven, "Conclusions."

Agencies in Society Today that Promote Evolutionary Ideas

Public schools— around the time of the Scopes "Monkey" trial in Dayton, TN in 1925, the teaching of biological evolution was banned in the public schools in many states. This situation continued until 1960 when a play called "Inherit the Wind" was made into a movie, starring Spencer Tracy, Gene Kelly and Frederic March. Phillip Johnson explains the far-reaching influence this movie had on Americans.

> The play is a fictionalized treatment of the "Scopes Trial" of 1925, the legendary courtroom confrontation in Tennessee over the teaching of evolution. Inherit the Wind is a masterpiece of propaganda, promoting a stereotype of the public debate about creation and evolution that gives all virtue and intelligence to the Darwinists. The play did not create the stereotype, but it presented it in the form of a powerful story that sticks in the minds of journalists, scientists and intellectuals generally. Inherit the Wind is a bitter attack upon Christianity, or at least the conservative Christianity that considers the Bible to be in some sense a reliable historical record. At the surface level the play is a smear, although it smears an acceptable target and hence is considered suitable for use in public schools... The real story of the Scopes trial is that the stereotype it promoted helped the Darwinists capture the power of the law, and they have since used the law to prevent other people from thinking independently.[21]

So, for five decades there has been an ongoing battle in the courts over the teaching of evolution and/or creation in the classroom, and the American Civil Liberties Union (ACLU) is leading the charge in favor of evolution. Consequently, in some states, the law allows the teaching of both creation and evolution. In other states only evolution is permitted.

However, exposure to the evolutionary view of life and origins is not confined to schoolrooms. Through television, films, books, activities and fashionable fads, men and women and children have been seduced into thinking that evolution is a fact. Because exposure to evolutionism is so widespread and pervasive in our society, it often passes unnoticed. A brief look at a few of the ways evolution has gained acceptance and popularity should be enough to awaken us to its inroads in our society.

Matrisciana and Oakland have convincingly shown how evolution has made these inroads into our thinking and culture through museums, television, dinosaurs and the *National Geographic* magazine.

Museums— The Smithsonian Institution in Washington, D.C. is America's premiere museum. There are so many fascinating things to see— Charles Lindbergh's *Spirit of St. Louis*, Mercury space capsules, dinosaurs and much more. However, many people don't realize that the Smithsonian Museum aggressively teaches evolutionism to millions of school children and adults. For example, the December 1981 *Smithsonian* magazine featured its "Tower of Time," a twenty-seven-foot mural in Dinosaur Hall at the National Museum of Natural History. This mural shows the 700,000,000-year span of life on earth from single cells to modern man. This statement, presented authoritatively as fact, gives no scientific evidence to support its conclusions. Indeed, there is none.[22]

In their book, *The Evolution Conspiracy*, Matrisciana and Oakland tell this revealing story:

> While the film crew was filming for "*The Evolution Conspiracy*, a Quantum Leap into the New Age" in the Smithsonian Institution, a leading paleontologist admitted off camera that the concept of the "Tower of Time" was basically speculation and without any scientific evidence. He further admitted that none of the theories about how life originated from non-living elements have concrete evidence to support them.[23]

Television— The media is a persuasive tool and able to effectively mold public opinion through one particularly powerful instrument— television— whether it be the regular network channels, ABC, NBC, CBS, or the Discovery and History channels. Several years ago, atheist astronomer Carl Sagan's thirteen-part TV series "Cosmos" captured the attention of millions of viewers. The cover of his book of the same title says,

> *Cosmos* is about science in its broadest human context, how science and civilization grew up together... [Do you notice anything amiss in that statement?] Sagan retraces the 15 billion years of cosmic evolution that have transformed matter into life and consciousness.[24]

Television programs like these war against an understanding of a Creator God who made man uniquely as man from the beginning and gave him an eternal destiny that the lower-life forms do not share.

Dinosaurs— In the 1980s the National Academy of Sciences and the National Association of Biology Teachers (NABT) teamed up to plan a strategy to suppress the teaching of creationism and advance the theory of evolutionism. One of the most appealing ways evolutionary theory was popularized was with the imaginative use of dinosaurs. It is one thing to acknowledge that dinosaurs existed as the fossil record shows; it is quite another to purposely indoctrinate children and adults worldwide with the idea that dinosaurs existed millions of years ago, before man, and that they prove the theory of evolution.

The *National Geographic* magazine, public parks and natural wonders across the country, films like "The Land Before Time" and "The Matrix", books, theme parks and PBS NOVA programs are all used to promote evolutionism while ignoring creationism. Little wonder children and adults think that evolution is the only reason we are here.

The *National Geographic* magazine has been in print for 125 years. Its issues are found in thousands of homes, libraries, places of business, reception rooms, hospitals and clinics. It has been popularized by its outstanding articles on scientific and natural discoveries and by its beautiful photographs of erupting volcanoes, wildlife in their natural habitat, the polar regions and many other subjects.

One significant historical observation is that the *National Geographic* magazine and Society were founded in 1888, just 29 years after the publishing of Darwin's first book, *Origin of the Species* in 1859. The proximity of these two happenings may be purely coincidental. But I think not! Over the years, I have read articles in the magazine that have been evolutionary in nature— that is, based on evolutionary assumptions— but with no basis in fact. This leads me to believe that the *National Geographic* magazine has had an evolutionary philosophy from the beginning.

The map of the world in Figure 2.1, showing the migration routes of mankind starting in East Africa and moving to the other continents,

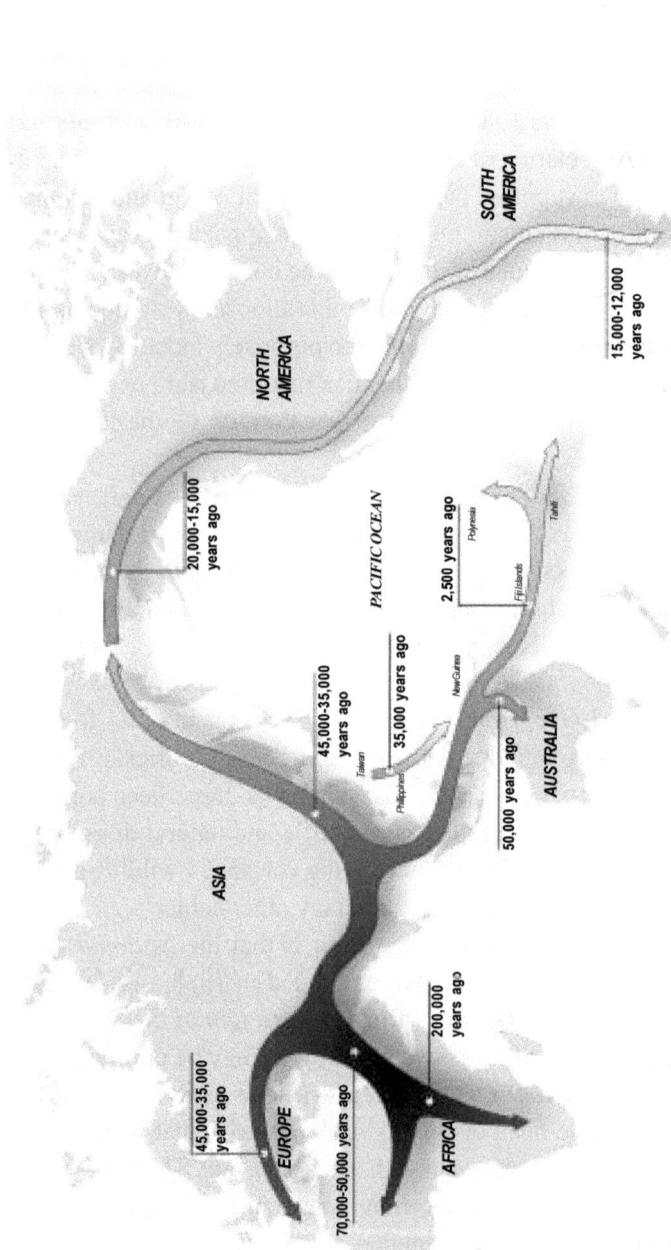

Figure 2.1: The "Global Journey:" The migration of modern humans according to Evolutionary Theory

According to evolutionary theory, "Once modern humans began their migration out of Africa some 60,000 years ago, they kept going until they had spread to all corners of the Earth. How far and fast they went depended on climate, the pressures of population, and the invention of boats and other technologies. Less tangible qualities also sped their footsteps: imagination, adaptability, and an innate curiosity about what lay over the next hill." (Adapted from *National Geographic Magazine*, Vol. 223, No. 1, January 2013. Used by permission.)

is based on a map that appeared in the 125th special anniversary issue of the *National Geographic* in January 2013, pp.48-49.

In Figure 2.2, notice the *contrast* between the two worldviews as to the origin and dispersion of the race.

Figure 2.2

The Evolutionary Worldview Global Journey	**The Biblical Worldview— Genesis**
1. This worldview begins with man and omits God completely.	1. The biblical worldview begins with God and moves to man. God is active throughout the narrative.
2. This view postulates that man first appeared in East Africa, implying that he descended from the ape some 200,000 years ago. (This is a theory with no evidence.)	2. According to the biblical account, man was created by God (1:26-27; 2:7) and placed in the Garden of Eden in Mesopotamia some 6,000 to 12,000 years ago (2:8, 15).
3. After some 70,000 to 50,000 years man began to migrate *on his own* to Arabia and Australia.	3. Man did not begin to migrate to other parts of the region or world until he was banished from the Garden of Eden because of his sin (3:24; 4:16). The biblical account says that it was God who did the scattering of the people at Babel over the face of the whole earth (11:1-10).
4. This account of man is very generic and impersonal. These persons are nameless. We don't know if they are real human beings or not.	4. This account is very specific and personal beginning with the first couple and their first three sons and their names. What follows are the generations of: Noah (6:9-9:28) Sons of Noah (10:1-11:9) Shem (11:10-26) Terah (11:27-25:11)

The Evolutionary Worldview	The Biblical Worldview— Genesis
	Ishmael (25:12-19) Isaac (25:19-35:29) Esau (36:1-43) Jacob (37:2-50:26)
5. The implied emphasis is on man descending from an animal.	5. The emphasis is on the special creation of man by God.
6. No mention of the creation of the universe or the earth, just an assumption that it may have existed for millions of years.	6. A distinct mention of the creation of the universe and the earth (1:1ff).
7. No mention of the creation of plant, animal and human life.	7. Ample mention of the plant (1:11-12), animal (1:20-25), and human life (1:26-27; 2:7, 20b-23).
8. No mention of the creation of matter, time and space.	8. The implication that matter, time and space were created out of nothing *(ex nihilo* – Heb. 11:3).
9. Apparently there is no purpose for man. Once modern humans appeared and began their migration out of Africa they kept going until they had spread to all the corners of the earth.	9. God blessed them and said to them, "Be fruitful and increase in number, fill the earth and subdue it. Rule over... (1:28). Man was made to have communion with his Maker (2:8-10).
10. No mention they disobeyed and fell into sin.	10. A clear mention of their disobedience and Fall into sin.
11. No mention of a prohibition not to eat of the tree of the knowledge of good and evil, nor of the consequences.	11. A clear mandate not to eat of the forbidden fruit and a warning of the consequences.

The Evolutionary Worldview	The Biblical Worldview—Genesis
12. No mention that Eve became the mother of all the living.	12. A clear statement about Eve's name and that she would become the mother of all the living (3:20).
13. No mention of the corruption of mankind and of God's judgment on the human race by a worldwide Flood.	13. A distinct mention of violence and corruption in the world, the destruction of everything on earth except Noah and his family by the worldwide Flood (6:1-8:22).
14. No mention of the beginning of civilization and societies, of nations and cultures, of human languages or of God's special people – Hebrews in the OT and Christians in the NT.	14. Specific mention of all four as follows: (a) civilization and nations (4:17b) (b) nations and cultures (10:1-32) (c) languages (11:1-9) (d) God's special people (11:10-50:26).
15. No mention of God's promises and plans to redeem sinful man.	15. The first mention of God's promises and plans to redeem sinful man (3:15).
16. No mention of the covenants God made with (a) Adam (b) Noah and (c) Abraham	16. Specific mention of these 3 covenants God made with important individuals: (a) Adam (3:15) *after* the Fall into sin (b) Noah (9:16) *after* the Flood (c) Abraham (12:2-3) *after* being called to leave his homeland of Ur of the Chaldees

Concluding Statements About these Two Worldviews

1. The evolutionary worldview moves solely on the *biological* and human anatomy level without any reference whatsoever to the Lord God, who made us in his own image.

2. The biblical worldview (account) moves on a higher moral and spiritual level, in addition to the biological and procreative level with specific reference to God's love, grace, mercy, compassion and judgment accompanying man's origin, nature, purpose and destiny from the dawn of creation throughout the primeval history of mankind which is fully revealed and recorded in Genesis 1-11.

3. Everyone knows, or should know, that the Garden of Eden located in ancient Mesopotamia (modern Iraq), and not East Africa, is the cradle of world civilization. This is abundantly authenticated not only by the biblical record (Gen. 2:10-14) but also by archaeological discoveries in Babylonia, the oldest culture in the Ancient Near East.

4. It is disappointing to see that the *National Geographic* magazine, which has been such an excellent instructive tool for informing millions of people about the natural world in which we live, rejects the real cradle of civilization in Mesopotamia and continues to perpetuate in the public mind that man, that was created in the image of the living God to glorify Him and enjoy His fellowship, originated in East Africa as a descendant of the chimp, with no missing link to prove this, and that it does this in defense of the theory of evolution.

5. Truly, this is one of the greatest frauds perpetrated on Western culture by Satan, the "god of this age" (2 Cor. 4:4) and the "prince of this world" (John 14:30), God's enemy and ours.

6. In this chapter we have presented the essence of naturalistic evolution that was begun by Charles Darwin in 1859 and early accepted by the intellectual and scientific communities in America.

7. In chapter three we will unfold the rise and spread of *Theistic Evolution*, a near relative of materialistic evolution, in Christian circles.

Endnotes

1. Howard Vos, *Exploring Church History* (Nashville: Thomas Nelson Publishers, 1994), 118.

2. R. E. Schofield, "The Lunar Society of Birmingham," *Scientific American* 247, June 1982.

3. Ian T. Taylor, *In the Minds of Men: Darwin and the New World Order* (Toronto: TYE Publishing, 1984), 55-57.

4. Ibid, 120.

5. Elizabeth A. Livingstone, ed., *The Concise Oxford Dictionary of the Christian Church* (Oxford: Oxford University Press, 1987), 531.

6. Taylor, *Minds of Men*, 67.

7. Caryl Matrisciana and Roger Oakland, *The Evolution Conspiracy* (Eugene, OR: Harvest House Publishers, 1991). 60.

8. Colin Brown, "The Ascent of Man," in *A Lion Handbook— The History of Christianity* (Herts, England: Lion Publishing, 1977), 528.

9. Gavin de Beer, Charles Darwin (London: Thomas Nelson and Sons Limited, 1963), 45.

10. Ibid, 307.

11. Charles Darwin, *The Autobiography of Charles Darwin, 1809-1882*, Appendix and notes by Nora Barlow, granddaughter of Charles Darwin (New York: W. W. Horton and Co., 1958), 85.

12. Quoted by Matrisciana and Oakland in *The Evolution Conspiracy*, 63.

13. Loc. cit.

14. Vergilius Ferm, ed., *An Encyclopedia of Religion* (New York: The Philosophical Library, Inc., 1945), 217.

15. Loc. cit.

16. Scott Huse, *The Collapse of Evolution* (Grand Rapid: Baker Book House, 1988), 90.

17. Earle E. Cairns, *Christianity Through the Centuries* (Grand Rapids: Zondervan Publishing House, 1958), 450.

18. Ibid, 450-451.

19. See my book, *Holiness for Earnest Christians*, pp. 65-66 and the secular non-biblical theories on sin, particularly no. 6.

20. Vos, op. cit. 120.

21. Phillip Johnson, *Defeating Darwinism by Opening Minds* (Downer's Grove, IL: InterVarsity Press, 1997), 25, 30, 32.

22. Matrisciana and Oakland, 27.

23. Loc. cit.

24. Ibid, 31.

3

The Rise and Spread of Theistic Evolution

Historic Christianity Becomes Divided over Theistic Evolution

IN CHAPTER TWO WE SPOKE OF the division evolution caused in Christian circles when Darwin's new theory broke on the world scene. The reaction of established religion in the latter part of the nineteenth century was threefold: some, namely, theological liberals, capitulated to the new "science" and turned their back on Christianity, including the Genesis accounts of Creation and the Flood, which by now had become the focused attack by Satan and the higher critics. Others repudiated the claims of science, namely, fundamentalists— that is, orthodox Christians— who believed in the authority, inerrancy and literal interpretation of the Scriptures. The majority worked out some sort of compromise between their faith and the new science. The liberal and conservative reactions to evolution in the latter part of the nineteenth century led, in part, to the Modernist/Fundamentalist controversies of the early twentieth century in America (1900-1930). The controversies were over the nature of the Scriptures, the nature of man, the nature of the world, the nature of Christ and man's need for salvation, and the nature of education in America. William James, who envisioned the inevitable perfection of humanity, and John Dewey, a humanist educator, had great influence on American education.[1]

The third reaction to evolution resulted in compromises with Darwinian thought in various forms within the Church from Darwin's time to the present. The most significant compromises and theories that arose were: 1) Uniformitarianism, 2) the Gap Theory, 3) the Day/Age Theory, 4) Theistic Evolution, and 5) the Big Bang Theory. These will be dealt with in greater detail in chapter ten, "Genesis 1 and 2:

What Does It Say and What Does It Mean?"

Theological Complications and Final Compromise

Let's step back for a moment and recall that in 1831 when the *HMS Beagle* sailed for the four corners of the globe, the discoveries of science and the acceptance of an incredibly well-ordered universe were still largely taken as evidence for the glorious design of an existing Creator God by the British and American people.

Unfortunately, as we have already noted, such belief and trust was gradually displaced by an atmosphere of doubt. This shadowed not only the scientific community, but the churches as well. Theological challenges to the Bible and questions began to surface everywhere. As Matrisciana and Oakland observe,

> Could one believe in Darwin's hypothesis and still hold to the account of creation in Genesis to be true? How should God's action as creator be perceived in relation to the evolutionary formation of new creatures? But the greatest question, the one that struck at the heart of the sacred Scriptures, was even more critical: If human beings evolved from lower animals to a higher state of intellectual and moral consciousness, *how could there be any place for the historic fall of man.*[2]

It would be impossible in this short space to trace all of the developments that took place as the Church began to accommodate the traditional beliefs in creation to the new ideas of biological evolution. However, the general path that unfolded can be understood by following the trend among church theologians with liberal tendencies. These men led the way toward conformity with the idea of evolution. Matrisciana and Oakland observe again,

> In their desperate desire to make the Bible seem credible and acceptable to the educated elite, many clergymen embraced the concept that the origin and history of life on planet Earth could be interpreted in evolutionary terms. By allowing long periods of time to exist within biblical chronology, and by emphasizing God's presence in nature [this is pantheistic] and the gradual progression of life toward the emergence of man, they pressed the Bible's account of origins into the evolutionary model.[3]

Having accepted the evolutionary expanded time frame for the history of the earth, theologians then searched for alternatives to the Genesis account of creation. The birth of a new view called "Theistic Evolution" was a compromise between the biblical and evolutionary worldviews of creation, which called into question the legitimacy of the Bible as a valid historical account of creation.

One important observation here is that these nineteenth century theologians doubted the validity of Genesis 1-11 and Genesis 1 and 2 in particular as an historical account of creation even though it had been divinely revealed to Adam and his descendants in the primeval period of history by the Creator Himself.

Another thing we need to keep our eye on as we move through the next chapters is the questions: Can evolution be accepted in Christian faith as Theistic Evolution? And, if it can be, where will this acceptance of Theistic Evolution by today's evangelical scholars and scientists lead them ultimately?

The Rise and Spread of Theistic Evolution

The most popular, attractive, and logical— or so it seemed to be at the time— of the compromise theories of creation mentioned above in the mind of many Christians, was Theistic Evolution because it gave a "nod" to the newly emerging scientific theory and, yet, kept God in the process of creation. It was the best of both worlds. Little by little, the concept of Theistic Evolution began to spread far and wide.

In America, two early strong proponents of Theistic Evolution were Asa Gray (1810-1888) and James Dana (1813-1895). When the theory of evolution first reached American soil Asa Gray, a Harvard profes-sor of botany, accepted it immediately and asserted firmly that it was not contrary to orthodox Christianity. He labeled himself as one "who is scientifically and in his own fashion, a Darwinian, philosophically a convinced theist, and religiously an acceptor of the 'creed commonly called the Nicene' as the exponent of the Christian faith."[4] Gray used his influence in every way he could to promote Darwin's ideas. Although he never went so far as to proclaim that the ape was the Adam of Genesis, a geology professor from Yale University by the name of James Dana did. Dana, who was a Christian, abandoned some of his former beliefs and became a convinced Darwinist after reading *Origin of the Species*. One of Dana's own claims was that he had made Yale a stronghold of evolutionary science, able to "correct false dogma [meaning for one thing the creation story in Genesis] in theological systems."[5]

In 1896, J. A. Zahm, a Roman Catholic scholar, defended Theistic Evolution very vigorously. He writes:

> From the foregoing pages, then, it is clear that far from being opposed to faith, theistic Evolution is, on the contrary, supported both by the declarations of Genesis and by the venerable philosophical and theological authorities of the Church.[6]

Zahm does not show how the declarations of Genesis support the theory of Theistic Evolution and one can only wonder if, given the orthodoxy of Roman Catholic theology, Theistic Evolution was supported by all of the venerable theological authorities of the Church. "The Roman Catholic Church," he affirms, "is not pledged to either evolution or special creation and *awaits the verdict of science* as to which was the *modus operandi of God*."[7] Science is now going to make this determination!

Among Protestant believers at the turn of the twentieth century is the name of W. N. Rice who strongly supported Theistic Evolution. In 1893 he writes:

> Now and then some theological Rip Van Winkle attempts the old Sinaitic thunders in denunciation of the essential atheism of Evolution; but his utterances are regarded by his brethren in the church not with sympathy, but with amusement and mortification. The curriculum of an orthodox theological seminary is hardly regarded as complete today without a course of lectures in the consistency of Evolution and theistic philosophy.[8]

Rice wrote the above for the Bibliotheca Sacra in an article entitled, "Twenty-five Years of Scientific Progress." Regarding this overblown optimistic statement of Rice, Bernard Ramm comments, "Theistic evolution had become so popular by the end of the nineteenth century that some writers voiced their opinion that the controversy was over with. They did not reckon with the Fundamentalist movement in the twentieth century and its dynamic, militant attack upon evolution and Theistic Evolution."[9]

But there were those Protestant scholars who, before the twentieth century, stepped forward in defense of the long-standing orthodox belief in creationism by God's fiat. One group of writers insisted that evolution itself is antichristian and therefore *Theistic Evolution is an impossible theory for a Christian to believe.* J. R. Straton writes:

> Those who try to reconcile these theories [of evolution] with the Christian system of truth assert that such is not the case... yet the definitions given... prove that God *is of necessity ruled out,* and that in favor of chance.[10]

D. J. Whitney asks if evolution can be theistic and he answers with a dogmatic negation.[11]

In a prize-winning essay read before the Victoria Institute in Great Britain, George M. Price, a staunch conservative, debates this very question. Price states his position in these very clear words:

> It is thus very evident that there is no similarity between the idea of Evolution and that of Creation; it is all contrast. The two terms are antonyms; they are mutually exclusive; no mind can entertain a belief in both at the same time; when one notion is believed, the other is thereby denied and repudiated.[12]

The battle lines are now drawn for the great debate in the twentieth century on this issue. In chapter four we will examine the timeline of the growing evangelical acceptance with regard to the question of *Theistic Evolution.*

Endnotes

1. See my book *Creation for Earnest Believers*, pp. 45-49, "Other Movements in America Leading to the Secular-Humanist Society Today."

2. Caryl Matrisciana and Roger Oakland, *The Evolution Conspiracy* (Eugene, OR: Harvest House Publishers, 1991), 67.

3. Ibid, 70.

4. Bernard Ramm, *The Christian View of Science and Scripture* (Grand Rapids: Wm. B. Eerdmans Publishing Company, 1955), 254-255.

5. Quoted by Matrisciana and Oakland in *The Evolution Conspiracy*, 72.

6. Quoted by Ramm in *The Christian View of Science and Scripture*, 282.

7. Ibid, 283.

8. Ibid, 284.

9. Footnote 36 in Ramm.

10. Quoted by Ramm in *The Christian View of Science and Scripture*, 281.

11. Loc. cit.

12. G. M. Price, "Revelation and Evolution: Can they be Harmonized?" *Journal of the Transactions of the Victoria Institute*, 57:169 (1925).

4

A Timeline of Growing Evangelical Acceptance of Theistic Evolution

A "Perfect" Solution

IN CHAPTER THREE WE NOTED briefly the history of the rise and spread of Theistic Evolution in England and America in particular, and the voices of some who opposed this theory on solid biblical grounds.

In time, however, this theory was seen by a growing number of Christians as a "perfect" scientific solution to the question of origins and created things without questioning evolutionary theory more carefully and also desiring to be "on board" and looking respectable by the scientific community.

The reasoning went like this. Perhaps, they justified, God had used natural processes as His method of creation, and had guided evolution to the final realization of man. Perhaps the creation of life as described in Genesis was nothing more than allegory or myth. Perhaps man needed to adopt a flexible point of view which allowed God a wide latitude in His method of creation.[1]

Early Signs of Compromise in Favor of Theistic Evolution

The above position of openness to Theistic Evolution and growing tolerance and acceptance of it by conservative scholars and churchmen continued throughout the twentieth century. The following are a few examples.

James Orr: In the Kerr Lectures of 1890-1891 he stated,

> On the general hypothesis of evolution, as applied to the organic world, I have nothing to say, except that, within certain limits, *it seems to me extremely probable, and supported by a large body of evidence.*[2]

Commenting on this Bernard Ramm says this:

Most amazing is that in *The Fundamentals* (IV, 91-104) Orr defends Theistic Evolution as interpreted by R. Otto's *Naturalism and Religion*. Orr calls it *creation from within*, and accepts a sudden mutation as in the case of the origin of man.[3]

It should be pointed out here that in his book, *The Christian View of Science and Scripture*, Ramm is very sympathetic to this new approach of evangelicals giving God broader latitude in describing His method of creation in Genesis 1. Though he himself is not a Theistic Evolutionist, he is a self-proclaimed progressive creationist.

Ruben A. Torrey was a staunch evangelical revivalist along with D. L. Moody, Christian educator, and first president of the Moody Bible Institute. In 1898 Torrey hinted that evolution might be true of animals but not of man. "Whatever truth there may be in the doctrine of evolution as applied within limits to the animal world, it breaks down when applied to man."[4] The question is: Why does evolution apply to animals? This is contrary to God's Word in Genesis 1:21, 24-25. Didn't he see this?

B. B. Warfield was one of the greatest Christian theologians of the nineteenth and early twentieth centuries. In 1911 Warfield makes a similar statement to Torrey's: if evolution be carefully guarded theologically it could pass as a tenable theory of the "divine procedure in creating men." Evolution cannot be a substitute for creation but "at best can supply only a theory of the method of divine providence."[5] Is not Warfield "cutting too much slack" for evolution?

It is interesting to note Francis Collins' words about B. B. Warfield in his book, *The Language of God*. He writes,

Benjamin Warfield, a conservative Protestant theologian in the late nineteenth and early twentieth century, was well aware of the need for believers to stand firm in the eternal truths of their faith, despite great social and scientific upheavals. Yet he saw also the need to celebrate discoveries about the natural world that God created.[6]

After paying homage to Warfield, Collins cites these words of Warfield written from the perspective of that day:

We must not, then, as Christians, assume an attitude of antagonism toward the truths of *reason*, or the truths of *philosophy*, or the truths of *science*, or the truths of *history*, or the truths of *criticism*. As children of the light, we must be careful to keep ourselves open to every ray of light. Let us, then, cultivate an attitude of courage as over against the investigations of the day. None should be more zealous in them than we. None should be more quick to discern truth in every field, more hospitable to receive it, more loyal to follow it, whithersoever it leads.[7]

Of course Collins, in citing these words of Warfield, is appealing to evangelical Christians today to be open to new scientific data and to embrace his version of Theistic Evolution set forth in his book and presented here in chapter five.

Albert Mohler counters both Warfield and Collins by shedding light on this openness of Warfield and other orthodox "Princeton theologians" to the novel idea in their day. He writes:

In their own way, even some among the honored and orthodox "Princeton Theologians" attempted to argue that there was no necessary conflict between Genesis and Darwin. They were so convinced of the power of empirical science and of the authority of Scripture that they were absolutely sure that the progress of science would eventually prove the truthfulness of the Bible.

What these theologians did not recognize was the naturalistic bent of modern science. The framers of modern evolutionary theory did not move toward an acknowledgement of divine causality. To the contrary, Darwin's central defenders today oppose even the idea known as "Intelligent Design." Their worldview is that of a sterile box filled only with naturalistic precepts.

Mohler goes on to add:

From the beginning of this conflict, there have been those who have attempted some form of accommodation with Darwinism. In its most common form this amounts to some version of "Theistic Evolution"— the idea that the evolutionary process is guided by God in order to accomplish His divine purpose.

Given the stakes in this public controversy, the attractiveness of Theistic Evolution becomes clear. The creation of a middle ground between Christianity and evolution would resolve a great cultural and intellectual conflict. Yet, the process of attempting to negotiate this new middle ground, it is the Bible and the entirety of Christian theology that gives way, not evolutionary theory. Theistic evolution is a biblical and theological disaster.[8]

A. H. Strong. In 1907 Strong wrote in the preface of his *System-*

atic Theology that he had accepted evolution into his thinking. He believes that the world had an evolutionary origin and progress. "Neither evolution nor the higher criticism has any terrors to one who regards them as part of Christ's educating process."[9] The Bible says the Holy Spirit is our guide into all truth (John 16:13).

A. R. Short, an outstanding British surgeon and evangelical Christian, wrote in 1942 that men before Adam might not have been men in the full sense of the term *man.*

> What sort of material the Creator used to make man, whether the dust of the earth directly, or the pre-existing body of a beast, we leave an open question.[10]

Why leave this an open question? Genesis 2:7 is clear. God used the dust of the ground. Short is sowing seeds of doubt. This is one of the sources for the theory of pre-Adamic men. Another source is C. S. Lewis in *The Problem of Pain.* Also see Collins, ppg. 208-209 and Richard N. Ostling, "The Search for the Historical Adam," *Christianity Today,* June 2011, pg. 27.

L. F. Gruber, an orthodox Lutheran, wrote in 1941 that the evolutionary origin of man would be permissible in Christianity if evolution be conceived as *God's method of creation.* "God as the Great Personal First Cause would be the Author or Creator at every point throughout the whole life-history."[11]

Albertus Pieters was a Reformed Church theologian. In 1943 in his *Notes on Genesis* admitted that one could believe in cosmic evolution or organic evolution and not compromise his Christianity. He himself does not accept evolution, but he is not willing to brand those who do as non-Christian. This raises an interesting question: How can a born again Christian, who holds to the authority and integrity of Scripture, accept an anti-biblical theory like evolution, which leaves the eternal Creator God out of the picture?

Pieters also says:

> If a Christian believer is inclined to yield as far as possible to the theory of organic evolution, he can hold that man's body was prepared by God through such a natural process, and that, when this process had reached a certain stage, God took one of the man-like brutes so produced, and made him the first human being, by endowing him with a human soul and a morally responsible nature... In such a

conception there is nothing contrary to the Bible.[12] [Question: Where is that found in the Genesis 1 and 2 account of the creation of man? This is pure speculation that cannot be proved empirically or biblically. It is an unwarranted concession.]

By way of summary, we can make these observations based on the views of Christian intellectuals who lived in the first half of the twentieth century:

1. There was a growing openness to and acceptance of Theistic Evolution by evangelical scholars and churchmen.

2. Most of them believed that God was the First Cause or Creator and that He used evolution as His method of creation.

3. Some also believed that the first human being appeared when God took one of the man-like brutes so produced and endowed him with a human soul and a morally responsible nature.

4. Some went so far as to believe that others could hold views like this and still be Christians.

5. It is surprising that great theologians of the past like B. B. Warfield argued that there was no necessary conflict between Genesis and Darwin. The only explanation is that they did not recognize the naturalistic bent of modern science. They had too much confidence that the evolving empirical science would eventually prove, or line up with, the truthfulness of the Bible.

6. However, the views of these otherwise orthodox Christian theologians are troubling and reflect the conscious or unconscious influence of Darwinian Evolution on them as they attempted to be open to the "new science" and yet be faithful to the central message of the Bible— redemption.

7. There can be no doubt but that the above views were expanded by the evangelical scholars and scientists of the latter half of the twentieth and early part of the twenty-first century.

Bernard Ramm's Groundbreaking Book (1955)

Ramm's book, *The Christian View of Science and Scripture*, is groundbreaking because, coming after the debates of the Modernist/ Fundamentalist Era in the twentieth century, this is the first attempt by a respected evangelical scholar to effect a harmony of science and Scripture, based on a rather thorough study of the fields of astronomy, geology, biology, and anthropology. The book is bold, provocative, honest and challenging. He contends,

> that there are two traditions in Bible and science both stemming from the developments of the nineteenth century... There is the ignoble tradition which has taken a most unwholesome attitude toward science, and has used arguments and procedures not in the better traditions of established scholarship. There has been and is a noble tradition in Bible and science, and this is the tradition of the great and learned evangelical Christians who have been patient, genuine, and kind and who have taken great care to learn the facts of science and Scripture. No better example can be found than that of J. W. Dawson but we would also include such men as John Pye Smith, Pratt, Dana, Hugh Miller, James Orr, Asa Gray, and Bettex.[13]

And yet, as we noted earlier, Gray and Dana did all they could in their day to promote Darwinian evolutionary ideas at Harvard and Yale.

When the Evangelical Book Club launched this book in 1955, one of the reviewers had this to say:

> Some will not be happy with his theory of Joshua's long day, or with his solution to Ahaz's shadow on the sun dial or with his advocacy of a local flood. His mythological-symbolical treatment of Genesis 1-3, his view of the antiquity of man, etc., will cause debate among Fundamentalists [and their evangelical successors]... While not adopting all his conclusions, we find the treatment the most satisfying, sane and intelligent of any we have read.[14]

The question in most evangelical reader's mind is this: How loyal has Ramm been to the divinely-revealed truth in Genesis 1 and 2 in light of the leeway he has given to certain scientific theories of geology and biology?

At the end of his book Ramm states his position in 1955 as an evangelical Christian and scholar.

> The writer is not a Theistic Evolutionist. He is a progressive creationist for he feels that in progressive creationism there is the best accounting for all the facts— biological, geological, and Biblical. He has friends who are fiat creationists and Theistic Evolutionists. Their respect for the Bible and their loyalty to Christ he admires. But progressive creationism is that theory of the relationship of God's works and

God's Holy Word which makes the most sense to the author— and upon what other basis can he make up his mind?[15]

Progressive Creationism is the belief in the instantaneous divine fiat acts of creation but that these divine acts were separated by long periods of time. This implies a belief in some form of evolution.

Ramm's book, while being an honest and noble attempt to bring about a harmony between science and Scripture, made it more acceptable for oncoming generations of evangelical scholars to promote deviant interpretations of Genesis 1 and 2, particularly Theistic Evolution, and still remain in the historic Christian fold.

Carl F. H. Henry's Historic Book (1946)

Carl F. H. Henry was arguably the greatest conservative, orthodox Protestant theologian and philosopher in America in the twentieth century. His remarkable life and accomplishments includes being professor at Northern Baptist Theological Seminary and later Fuller Theological Seminary, being the first editor of *Christianity Today*, the flagship of the newly emerging evangelical cause in America, founded by Billy Graham, and a prolific writer and author of more than ten books. Perhaps one of the most important books was *Remaking the Modern Mind,* written in 1946. Writing at the close of World War II, he and others were calling for a remaking of the modern philosophical mind that was based on the premises that nature is the ultimate reality and that man is only an animal. Henry contends correctly that over the last 350 years the modern mind had been shaped by rationalism, empiricism, and positivism that left no room for supernatural revelation and the belief that man stands in unique relation to His Creator and to the objective, eternal moral order. Henry goes on to assert that naturalism made the break with the traditional, biblical view of God, man, and the universe of the pre-modern era complete, discarding the notion of a projected Absolute as vigorously as it discarded the idea of a divine revelation. With Hume, Comte and finally Dewey, this naturalistic philosophy captured the western mind.

A century ago (1859), modern philosophy took for its bride the scientific theory of evolution; in the course of the long wedlock there was occasion for many offspring, especially within the pattern of activism.

But the predominant family resemblance… was everywhere apparent: the world of nature, and man as its most important product, became the key to all reality. Since the turn of the century the almost undisputed emphasis in the great universities of the Occident has been that not of the Judaistic-Christian tradition, nor of idealism whether of the classic Greek or the specifically modern variety, but of *naturalism*.[16]

Later on in the book, when discussing the influence of Darwin's evolutionary thinking on the western world, Henry reluctantly admits that "science had become not the handmaid of theology [which should be the case] but of evolutionary metaphysics."[17] This naturalistic view of man is so dominant in our culture today that it has begun to taint even evangelical theology in some circles.

The Appearance of Ecclesiastical Organizations

To blunt the power of the modernistic (liberal) Federal Council of Churches, organized in 1908 and reorganized in 1950 as the National Council of Churches in America, there were organized the American Council of Churches by Carl McIntyre in 1941 and the National Association of Evangelicals (NAE) by a group of leading evangelicals in 1942. Both of these organizations represent the philosophy of Christian theism, and are loyal to the tenets of sacred Scripture. However, the American Council of Churches is an outgrowth of the old fundamentalism, whereas the NAE represents the new evangelicalism.

The objective of the NAE is to make just one conservative and united evangelical voice heard in important subjects such as missions, education, evangelism, social action and world relief for the nations of the Third World. In 1944 a group of ten to twelve professors and leaders met to discuss topics like theism, revelation, and the substitutionary satisfaction of Christ's atonement, which were relevant topics in that day. Surely the leaders of the newly formed NAE were also concerned about the dangers of materialistic evolution but in their minds it had not yet gained the attention it would in a few years. This would be because of the growing dominance of evolutionary thinking in education, in politics, in the media, and in social practice, and with the founding of the Institute for Creation Research and the creationist movement, and the growing acceptance of Theistic Evolution.

The Appearance of Scientific Affiliations

Shortly after the founding of the National Association of Evangelicals in 1942, the American Scientific Affiliation (ASA) was created in 1944 with a membership of some 800 evangelical scientists. The original purpose of this organization was to bring together Christian scientists who were committed to the belief that the Bible is the inspired Word of God and who, through their meetings and journal, could discuss many thoughtful proposals of a pathway toward harmony between science and faith. Bernard Ramm sought a harmony of science with Scripture. Here the harmony is between science and faith, that is, "serious believers in God."[18] This is an important distinction and one that will lead to a further erosion of the authority and integrity of the Scriptures.

In the beginning, this seemed to be a noble goal, but with the passing of time the organization became dominated by a belief in Theistic Evolution as Dr. Frank Cassel, head of the North Dakota State University Department of Zoology and then-president of the ASA, admitted in an article entitled, "The Evolution of Evangelical Thinking on Evolution" in the 1959 ASA Journal:

> Thus, in fifteen years, we have seen develop in A.S.A. a spectrum of belief in evolution that would have shocked all of us at the inception of our organization. Many still reserve judgment but few, I believe are able to meet Dr. Mixter's challenge of, "Show me a better explanation." Some may see in this developing view the demise of our organization, but it seems to me that we only now are ready to move into the field of real potential of contribution— that is releasing Truth from the restrictions we have been prone to place upon it, we can really view it in the true fullness which the Christian perspective gives us.[19]

Commenting on this departure from orthodox creation truth, Henry Morris observes:

> Dr. Richard Mixter, whom he [Cassel] cites and who was head of the Zoology Department at Wheaton College, had likewise swung largely to the evolutionary viewpoint in recent years. He [Mixter] says: "Genesis 1 is designed to tell who is the Creator, and not necessarily how the full process of creation was accomplished."[20]

Yet scientists are more than ready to offer their evolutionary theories to explain the how.

Morris is quick to point out that

This is a very popular rhetorical device of Theistic Evolutionists. But if the only purpose of the Creation account is to tell us that God is the Creator, then what is the value of the rest of the account? Why does not the record simply stop at the end of Genesis 1:1, which gives us this information quite adequately.[21]

Then writing prophetically, Morris says, "Neither the sincerity nor the good intentions of these brethren is questioned, but the writer strongly believes that the long-range results of these defections will prove tragic."[22] [This was written in 1967.] Present membership in ASA numbers 1600 scientists and scholars.

The most recent personality to enter the creation/evolution controversy and debate is geneticist and researcher Dr. Francis Collins. He captured the headlines in the summer of 2006 with the release of his controversial book *The Language of God: A Scientist Presents Evidence for Belief*, and his subsequent invitation to debate atheist Richard Dawkins on origins in September 2006.[23]

In late 2007, Collins launched the San Diego-based BioLogos Foundation which has as its purpose "to promote Theistic Evolution, especially among evangelicals. He sought not only to embrace what he considers to be the best evidence, but also to bolster Christian credibility among people who are knowledgeable about mainstream scientific thinking. This initiative has won endorsements from both scientists and such evangelical figures as authors Os Guiness and Philip Yancy, *Book and Culture* editor John Wilson, and retiring Gordon College President R. Judson Carlberg."[24]

In 2009, President Obama chose Francis Collins to be the director of the National Institutes of Health (NIH) in Bethesda, Maryland. Collins, one of the most eminent scientists ever to identify as an evangelical Christian, staunchly defends Darwinian Evolution even as he insists on God as the Creator. And he now stands at the epicenter of a dispute that increasingly agitates fellow believers. At issue: the traditional tenet (as summarized in Wheaton College's mandatory credo) that "God directly created Adam and Eve, the historical parents of the entire human race."[25]

There are two reasons or assertions (beliefs) reported in Collins' book *The Language of God* that are causing this dispute with fellow believers.

First, he believes that there are "scientific indications that anatomically modern humans emerged from primate ancestors perhaps 100,000 years ago— long before the apparent Genesis time frame— and originated with a population that numbered something like 10,000, not two individuals. Instead of the traditional belief in the specially created man and woman of Eden who were biologically different from all other creatures, Collins mused, might Genesis be presenting 'a poetic and powerful allegory' about God endowing humanity with a spiritual and moral nature? 'Both options are intellectually tenable,' he concluded."[26]

Second is the genetic argument. Dennis Venema, the BioLogos senior fellow for science and the biology chairman at Trinity Western University, claims that

> the chimp genome (total genetic heredity encoded in DNA), which was fully mapped by 2005, displays "near identity" with the human genome as detailed by Collin's team, with a 95 to 99 percent match depending on what factors are included... The cumulative evidence, Venema concludes, shows that "humans are not biologically independent, *de novo* creations, but share common ancestry" with prior primate species. (Many biologists estimate that the biological branches separated from that common ancestor some 5 or 6 million years ago.)[27]

Yet, Fazale Rana, the vice president for research with Reasons to Believe, a ministry that champions old earth creationism, questions the ninety-five to ninety-nine percent figures, but asserts "that in any case common sense tells us 'these types of genetic comparisons are meaningless' because they do not explain the 'fundamental biological and behavioral differences' between chimps and humans. Rana also says close genetic similarity does not require shared ancestry."[28]

Nevertheless, evangelical scientists, scholars and biology professors in a number of evangelical Christian colleges, like Calvin College, Eastern Nazarene College, Gordon College, Wheaton College, Westmont College, and Trinity Western University, have "bought into" these new scientific ideas at the risk of losing their academic positions and tenure, which has happened in several instances.

The situation is so grave that in 2011 Michael Cromartie, the evangelicalism expert at Washington's Ethics and Public Policy Center, wrote that he sees "high stakes, calling the new thinking an 'urgent' and 'potentially paradigm-shifting' development with 'huge theological implications… How this gets settled is extremely important.'"[29]

What May Be At Stake

What is at stake here is not only the authority of the Bible but also the future of the gospel. The stakes are so high in this dispute that I want to include this quote from *Christianity Today:*

> Foundational confessions of faith from the Protestant Reformation assume a historical Adam, and official Roman Catholicism defined this teaching at the 1546 Council of Trent… The broader public is intrigued, more so than by many other biblical topics; a 2005 Gallup Poll found that 40 percent of Americans think the various competing concepts of human origins matter "a great deal." So, is the Adam and Eve question destined to become a groundbreaking science-and-Scripture dispute, a 21st century equivalent of the once disturbing proof that the Earth orbits the sun? The potential is certainly there: the emerging science could be seen to challenge not only what Genesis records about the creation of humanity but the species' unique status as bearing the "image of God," the Christian doctrine on original sin and the Fall, the genealogy of Jesus in the Gospel of Luke, Paul's teaching that links the historical Adam with redemption through Christ (Rom. 5:12-19; 1 Cor. 15:20-23, 42-49; and his speech in Acts 17).[30]

Collins and his colleagues dismiss the 1) young earth, 2) old earth, 3) intelligent design in favor of Theistic Evolution, which affirms that the biblical God was the Creator of all earthly organisms, humanity included, and used as his method the standard evolutionary scenario of gradual natural selection among genetic mutations across eons. A non-random Internet survey of teachers at evangelical seminaries in 2009 showed that forty-six percent accept that concept. Giberson estimates that "the overwhelming number in biology departments at Christian colleges would be fine with this," though a 2005 survey found that only twenty-seven percent identified as evolutionary creationists. In a mail survey of ASA scientists last year, sixty-six percent of respondents affirmed that "*Homo sapiens* evolved through natural processes from ancestral forms in common with primates," while ninety percent agreed that the Earth is some 4.6 billion years old.[31]

But there is more. In November 2010, BioLogos held a workshop at New York City's Harvard Club where church leaders and Christians in science deliberated on evolution, creation, and Adam. The meeting issued an accord that endorsed Theistic Evolution and affirmed "without reservation both the authority of the Bible and the integrity of science" as two paths of divine revelation. The paper declared that "several options" can achieve a synthesis between Scripture and science, "including some which involve a historical couple, Adam and Eve." Participants in the discussion that produced the statement included Francis Collins, Michael Cromartie, Peter Enns, Darrel Falk, Karl Giberson, Os Guiness, Randall Isaac, and Philip Yancey.[32] We are witnessing how far compromise on basic biblical and divinely-revealed truths can lead some Christians, in this case, evangelical intellectuals.

Despite these disturbing views of some evangelicals, Gallup surveys report that at least forty percent of the general public believes in "young earth" creationism.[33] Obviously, not everyone in the evangelical fold is "on board" with these latest scientific proposals.

Another participant in the BioLogos workshop was Manhattan pastor Tim Keller. Even though he is an advocate for Theistic Evolution, he did present a theological connection that cannot be surrendered, namely,

> Paul most definitely wanted to teach us that Adam and Eve were real historical figures. When you refuse to take a biblical author literally when he clearly wants you to do so, you have moved away from the traditional understanding of the biblical authority. If Adam doesn't exist, Paul's whole argument— that both sin and grace work convenantally— falls apart. You can't say that "Paul was a man of his time" but we can accept this basic teaching about Adam. If you don't believe what he believes about Adam, you are denying the core of Paul's teaching.[34]

South Carolina pastor Richard Phillips, a blogger with the Alliance of Confessing Evangelicals and chair of the Philadelphia Conference on Reformed Theology,

> sees serious doctrinal danger if the historical Adam disappears. Can the Bible's theology be true if the historical events on which the theology is based are false? he asks. If science trumps Scripture, what

does this mean for the virgin birth of Jesus, or his miracles, or his resurrection? The hermeneutics [manner of interpreting the Scriptures] behind theistic evolution are a Trojan horse, that once inside our gates, must cause the entire fortress of Christian belief to fall.[35]

In chapter five we will present a biblical response to and critique of Francis Collins' recent book, *The Language of God*. Stay tuned.

Endnotes

1. Caryl Matrisciana and Roger Oakland, *The Evolution Conspiracy* (Eugene, OR: Harvest House Publishers, 1991), 72.

2. Bernard Ramm, *The Christian View of Science and Scripture* (Grand Rapids: Wm. B. Eerdmans Publishing Company, 1955), 286.

3. Loc. cit.

4. Ibid, 287.

5. Loc. cit.

6. Francis Collins, *The Language of God* (New York: Free Press, 2006), 179.

7. Loc. cit.

8. R. Albert Mohler, "The New Shape of the Debate," *Southern Seminary* (Winter 2011, Vol. 79, No. 1), 24-25.

9. Ramm, op. cit. 287.

10. Ibid, 288.

11. Ibid, 287.

12. Ibid, 288.

13. Ramm, Preface.

14. Reviewers' comments on Ramm's book.

15. Ramm, op. cit., 293.

16. Carl F. H. Henry, *Remaking the Modern Mind* (Grand Rapids: Wm. B. Eerdmans Publishing Company, 1946), 23-24.

17. Ibid, 149.

18. Collins, op. cit., 198.

19. Henry M. Morris, *Studies in the Bible and Science* (Philadelphia: Presbyterian and Reformed Publishing Co., 1967), 89-90.

20. Ibid, 90.

21. Loc. cit.

22. Loc. cit.

23. David Van Biema, "God vs. Science," *Time* (November 13, 2006). 48-55.

24. Richard Ostling, "The Search for the Historical Adam," *Christianity Today*, June 2011, 25.

25. Ibid, 23.

26. Ibid. 24.

27. Ibid, 25.

28. Loc. cit.

29. Ibid, 24.

30. Loc. cit.

31. Ibid, 25.

32. Ibid, 27.

33. Ibid 24.

34. Ibid, 27.

35. Loc. cit.

5

A Biblical Response to and Critique of Francis Collins' Book, The Language of God

Biographical Background

D R. FRANCIS COLLINS' RISE to prominence began when, as a young college student, he signed up for a course in biochemistry investigating the life sciences (pg. 17). This led him to leave the program he was in and enter med school at the University of North Carolina. A series of lectures on medical genetics, illustrated by patients with sickle cell anemia, galactosemia and Down's syndrome who were brought to class, showed him his future (pg. 18).

At this point in his life (1973), he had no idea of one day being a part of one of the most historic undertakings of humankind— the Human Genome Project. Over the next few years, through his contact with sick and dying patients and his reading of C. S. Lewis' *Mere Christianity,* he said he was moved to leave his atheism, and later agnosticism, into a belief in God and a personal relationship with Jesus Christ as his Savior (ppg. 21-31).

When he was a research fellow in genetics at Yale University in the early 1980s the decoding of DNA had already begun, but it was an arduous undertaking (pg. 109). For the benefit of the reader, DNA is a nucleoprotein which assures that living organisms will reproduce after their kind. More simply, DNA are long chemical strands (double helix) in living cells which contain information that specifies, procreates and controls all of life. Orthodox Creationists believe that God did this at the mature creation of all living forms (see Genesis 1:20-22, 24-25; 2:7, 20b-21). Another definition is in order here. A genome is one haploid set of chromosomes with the genes they contain.

While at Yale he learned "that a few visionary scientists had begun to discuss the possibility of determining the DNA sequence of the entire human genome, estimated to be about 3 billion base pairs in length and all of the genes that determine human heredity... We knew relatively little then about what the genes might contain. No one had actually seen the chemical bases of an individual human gene under the microscope (they were too tiny)" (ppg. 110-111).

Finally, as a physician and a junior researcher, Collins "decided to join the ranks of those who were undertaking an organized program to sequence the human genome — Human Genome Project. Despite all the uncertainties, there was no question how valuable a complete genome sequence would be. Hiding in this vast instruction book would be the parts list of human biology, as well as clues to a long list of diseases that we understand poorly and treat ineffectively." (pg. 111).

To shorten this rather long and detailed journey as a rising researcher, let me summarize the rest of the story with these key dates.

1. By 1996, "we were ready to start plotting the actual large-scale sequencing of the human genome."

2. By 2000, "we announced a first draft of the human genome instruction book which had been determined." "The language of God was revealed" (pg. 122).

   Observation 1. This statement becomes the basis for BioLogos workshop participants in 2010 to declare that science is a second path of divine revelation. Observation 2. The uncovering of the human genome is a scientific discovery— grand and noble to be sure— but not a divine revelation in the theological sense.

3. By 2003, all of the goals of the Human Genome Project had been met and there was great rejoicing by Collins, the project manager, and the more than 2000 scientists from around the world who had accomplished this remarkable feat, "that I [Collins] believe will be seen a thousand years from now as one of the major achievements of humankind" (pg. 122).

4. By 2005, the chimpanzee genome was fully mapped. (For the fully mapped genome of other living creatures see pg. 128.)

5. In 2006, Collins writes his book, *The Language of God,* describing in great detail this remarkable discovery and what he believes its significance to be.

6. In 2007, Collins launched the San Diego-based BioLogos Foundation to promote Theistic Evolution, especially among evangelicals.

7. In 2009, "Collins won unanimous US Senate confirmation, to be the director of the National Institutes of Health (NIH), thanks to sterling achievements in biological research and leadership of NIH's human genome research."[1]

However, "Collins, one of the most eminent scientists ever to identify as an evangelical Christian, staunchly defends Darwinian evolution even as he insists on God as the Creator. And he now stands at the epicenter of a dispute that increasingly agitates fellow believers. At issue: the traditional tenet (as summarized in Wheaton College's mandatory credo) that 'God directly created Adam and Eve, the historical parents of the entire human race.'"[2]

Why is he at the epicenter of this dispute with fellow believers? Because of certain beliefs he has stated in his book, which are contrary to Genesis 1 and 2.

Space does not permit a full review of his logic and his positions. So, this review will be limited to the two most important chapters in his book: Chapter five, "Deciphering God's Instruction Book," and Chapter ten, "Option 4: BioLogos (Science and Faith in Harmony)".

Inasmuch as Dr. Collins has chosen to challenge traditional Creationists on biological grounds, I shall challenge him on biblical and theological grounds. In this chapter I propose to set forth:

1. The goal, central question, argument, and methodology of his book, and
2. a critique of chapter five with rebuttal.

In chapter six, I will:

1. Give a critique of chapter ten with rebuttal,
2. Draw a sharp contrast between Francis Collins and a true creationist, and
3. Conclude with some closing observations and remarks.

Opening Remarks

Goal of the book, The Language of God. "The goal of this book is to explore a pathway toward a sober and intellectually honest integration of both the scientific and spiritual perspectives and views" (pg. 6).

Collins recognizes that science, interpreted specifically as evolution or some form of it, and the Christian faith, interpreted more specifically as the creation of the world and man by God in a literal six-day period, have been at odds ever since Charles Darwin launched his theory of evolution in 1859.

Central question of this book. "In this modern era of cosmology, evolution, and the human genome, is there still the possibility of a richly satisfying harmony between the scientific and spiritual [translated, "belief in God's existence and Christ as Savior"] worldviews? I answer with a resounding yes!" (pg. 6)

The problem here is that, from the outset, his view of the scientific worldview, for the most part, is naturalistic, and it is cloaked in the general term "scientific."

> In my view, there is no conflict in being a rigorous scientist and a person who believes in a God who takes a personal interest in each one of us. Science's domain is to explore nature. God's domain is in the spiritual world, a realm not possible to explore with the tools and language of science. [This is a clear demarcation.] It [God's domain] must be examined with the heart, the mind, and the soul— and the mind must find a way to embrace both realms (pg. 6).

The embracing of the scientific realm depends on what science has to offer the thinking person. The believer cannot accept a scientific statement, for example, that is not in harmony with revealed truth in Genesis 1 and 2, or the Bible in general.

His argument.

> I will argue that these perspectives [scientific and spiritual] not only *can exist within one person*, but can do so in a fashion that enriches and enlightens the human experience. Science is the only reliable way to understand the natural world, and its tools when properly utilized can generate profound insights into material existence. But science is powerless to answer questions such as, "Why did the universe come into being?" "What is the meaning of human existence?" "What happens after we die?" One of the strongest motivations of humankind is to seek answers to profound questions, and we need to bring all the power of both the scientific and spiritual perspectives to bear on understanding what is both seen and unseen. (pg. 96)

How does the power of science bring answers to the profound

metaphysical questions stated above?

His methodology. Collins' underlying objective throughout the book seems to be to convince Christians (believers in the Word of God) that they can:

1. hold fast to the *concept* of God as Creator;
2. hold fast to the truths of the Bible;
3. hold fast to the conclusion that science offers no answers to the most pressing questions of human existence, and
4. hold fast to the certainty that the claims of atheistic materialism must be resisted (pg. 178)

and, at the same time, believe the unbiblical, non-traditional, and as-yet-unproven scientific assumptions he sets forth in his book concerning the origin and age of the universe and the evolutionary origin and descent of man. As a serious believer and a serious biologist, he is asking the reader to *compromise* his belief in what the Genesis account really says about the origin of the world and human beings. He attempts to do this

- by showing his personal journey from atheism to agnosticism to a belief in God's existence and to his acceptance of Jesus Christ as his Savior from sin, which is not unlike C. S. Lewis' journey from atheism to belief in God, whose ideas he cites in support of his so-called mediating position between science and faith;
- by showing that what he has said about science and faith can coexist within one person (pg. 6);
- by showing the needlessness of warfare between these two worldviews of science and faith; [The "needlessness of this warfare" is a fallacy. This warfare becomes a necessity when scientific views of man's origin, like those expressed in this book, impinge on the integrity and truthfulness of Genesis 1 and 2]; and
- by presenting four (4) possible responses to the contentious interaction between the theory of evolution and faith in God [which should read, "between the theory of evolution and revealed Scripture."]

The four responses are:

Option 1: Atheism and Agnosticism (When Science Trumps Faith)
Collins, for all intents and purposes, rules this option out because

atheists do not believe in the existence of God and agnostics don't know whether God exists or not.

Option 2: Creationism (When Faith Trumps Science)

Collins denigrates creationists by contending that Young Earth Creationism and modern science are incompatible and by taking issue with the belief that Adam and Eve were real historical figures, created by God from dust in the Garden of Eden and not descended from other creatures. Then he cites Benjamin Warfield, that "as Christians we should not assume an attitude of antagonism toward the truths of reason, or the truths of science… but we must be careful to keep ourselves open to every ray of light."

It is unlikely that Warfield would have taken this position, if he were alive today, in the light of the false scientific evidence about the origin of the universe and man that is being perpetrated on believers in our day.

Option 3: Intelligent Design (When Science Needs Divine Help)

First, Collins describes the three fundamental propositions of Intelligent Design (ID), and then he proceeds to debunk it as a legitimate scientific theory by showing both the scientific and theological (?) objections to ID.

At the end of this chapter he asks, "So is the search for harmony between science and faith hopeless?" To the believer and the scientist alike, I say there is a clear, compelling, and intellectually satisfying solution to this search for truth.

Then he proceeds to give his option that supposedly has the above qualities.

Option 4: BioLogos (Science and Faith in Harmony)

For Collins, BioLogos is really Theistic Evolution, which incorporates all of the fundamental assumptions of Darwinian naturalism.

After stating that Theistic Evolution is the dominant position of serious biologists who are also serious believers, he proceeds to lay down the six (6) premises of Theistic Evolution, which will be critiqued in chapter six.

A Critique of The Language of God,

Chapter Five, "Deciphering God's Instruction Book: The Lessons of the Human Genome"

Core Idea of the Chapter. In this chapter Collins describes the process that led the team of 2000 research scientists from around the world to fully map the human genome which contains all of the genes that determine human heredity. He calls this accumulated data, "God's instruction book." "This book was written in the DNA language by which God spoke life into being… The Language of God was revealed" (ppg. 123, 122).

A clarification about this wording. Collins seems to be using the words *revealed* and *revelation* in the sense of "something that is revealed by God to man."[3] This is the secondary meaning of the term. The primary meaning is "an act of revealing or communicating divine truth."[4] The primary meaning of revelation is biblical and theological rather than scientific. Noah Webster (1758-1843), a Christian and American lexicographer who lived at the time of the founding of our nation, was right in making this distinction. Revelation in the true sense of the word has to do with the communication of divine truth from God to man about himself, and about man, his origin, his fall, his sin, his redemption through Jesus Christ, and his afterlife. It was this confusion of terms that may have led the participants in the 2010 BioLogos workshop to declare that the "integrity of science" is a second path of divine revelation.[5]

Fully mapping the human genome was truly an outstanding scientific discovery— one which "will be seen a thousand years from now as one of the major achievements of humankind" (pg. 122). But it must be remembered that it was God who enabled the human team to accomplish this remarkable discovery for the benefit of humanity. Just like God had led Lord Kelvin to discover the absolute temperature scale, or William Harvey to discover the circulation of blood in the human body, or Louis Pasteur to discover the law of biogenesis. Sir Isaac Newton, the great English scientist, believed that his scientific *discoveries* were communicated to him by the Holy Spirit, and regarded the understanding of Scripture as more important than his scientific work.[6] This is to emphasize the distinction between the revelation of divine truth and scientific discovery.

In another place, Collins says, "For me, as a believer, the uncovering of the human genome sequence held additional significance. This book was written in the DNA language by which God spoke life into being" (pg. 123).

Rebuttal. Perhaps Dr. Collins is using "book," "language" and "spoke" in a metaphorical sense. But in so doing, he only adds more confusion to this subject, and he tends to blur the distinction between the spiritual and the scientific. For example, the question is often asked: Which came first, the chicken or the egg? It is obvious that the chicken came first. So let me ask this: Which came first, Divine revelation about origins or science? Again the answer is obvious. Divine revelation, the written Word of God, came first. Science followed much, much later. So, my next question is: Are these two books— the human genome "instruction book" and the Bible— of equal value and weight? I think not. The hymn writer was inspired to pen these words:

> *Holy Bible, book divine,*
> *Precious treasure, thou art mine;*
> *Mine to tell me whence I came;*
> *Mine to teach me what I am.*[7]

The human genome "book" only tells me how I function. The divine book tells me where I came from and whose I am— a son bearing the imprint of the Father, bearing His image in wisdom, righteousness, love, and holiness.

The other thing that is in question here is this: Did life come into being by the DNA language written in this "book," or directly by God breathing life into the first human being, Adam? The first way sounds like a mechanism of evolution. The second way is the revelational intent of the writers of the Scriptures— the Divine author (the Holy Spirit) and the human writers guided by the Spirit as they wrote. In Psalm 33:6, 9 we read,

> By the word of the Lord were the heavens made
> Their starry host by the breath of his mouth.
> For he *spoke* and it came to be;

He commanded, and it stood firm. (NIV)

The English Standard Version (ESV) commentary of this passage reads like this:

> *God's Word...* is spoken by the same God who made everything (vv. 6-9). Verses 6-9 echo the creation account (Gen. 1:1-2:3), where each time God *spoke,* what he *commanded* produced its effect. The Septuagint Greek of Ps. 33:6 with the *word* (Gr. Logos) as the means [not the mechanism] of creation, probably lies behind John 1:3; the Word came to be seen as a personal agent, which John identifies as Christ himself (cf. John 1:14).[8]

What is Collins Trying to Prove in This Chapter?

What is Dr. Collins attempting to prove or decipher by the full mapping of human genome, which he calls "God's instruction book?" From a careful study of chapter five it seems that Collins has several objectives in mind.

1. To show that biologically, human beings are a part of the animal kingdom, albeit a special class of animals.

Collins writes, "Applying evolutionary science to Sticklebacks may be one thing, but what about ourselves? Since Darwin's time, people of many worldviews have been motivated to understand how the revelations [really, "discoveries"] about biology and evolution apply to that special class of animals, human beings. The study of genomes [in animals and humans] leads inexorably to the conclusion that we humans share a common ancestor with other living things" (ppg. 133-134). The tree of life shown on page 128 depicts the closeness or distance of relationships between different mammalian species.

Later on, Collins admits, "already... evolution has been the source of great discomfort in the religious community over the past 150 years, and that resistance shows no sign of lessening. Yet believers would be well advised to look carefully at the overwhelming weight of scientific data supporting this view of the relatedness of all living things, including ourselves" (pg. 141).

Rebuttal. At a distance men and apes do look alike; so do typewriters and computers. Does this mean that computers evolved from typewriters? We might say yes, if we include their common designer;

but typewriters don't evolve, they wear out. A typewriter can't produce a computer any more than a light bulb can make a star. On the contrary, machines rust and bulbs burn out. Where then does the notion come from that man evolved from apes?

Similarity on the physical level. Admittedly, we know that they share common structural and anatomical features.[9]

Similarity on the psychical level. Human *experience* tells us that animals share with humans certain psychical traits on a very elementary level. Take *feeling,* for example. Some animals have some degree of feelings as humans do. They are somewhat sensitive to heat and cold, and certainly to pain and abuse. This is why we have the Society for the Protection of Cruelty to Animals (SPCA). This is part of our stewardship. Man, as God's ambassador and agent on earth, is to rule or have dominion (Gen. 1:28-29) over his creation, as God rules all things— with wisdom, love, holiness and righteousness. This is the functional component of our bearing the Creator's image.

A second psychical trait man has with some animals is that of *expectation* and *hope.*

A third area of similarity between some animals like the chimp and man is in the area of *reasoning/thinking.*[10]

One question that comes to mind: Do we have any indication that man is intrinsically different than the animals? Is there anything that we can point to that shows a fundamental separation between man and the animal kingdom? Research scientist and professional engineer Robert Gange offers this analogy by way of contrast.

> One thing we might observe is that whereas man lands on the moon, monkeys land in trees. Whereas chimps use their arms to travel from tree to tree, man uses jets to travel from continent to continent. And whereas apes beat their chests to send noise through the jungle, man orbits satellites to send radio signals through space.[11]

The main thing this all suggests is that man is not just another animal on a higher plane, but rather a *separate* level far above the animals. So, if man is on a separate level far above the animals, what makes man unique among all created beings? Does man reflect a supernatural image? Theologian Eugene Carpenter would say yes, he does reflect that image. In his essay on "Biblical Cosmology" he writes with deep feeling,

> Man is, according to Genesis 1 and 2, not a product of the forces of nature, but a creation of God who acts in order to bring man into existence. Man indeed shares on one level physical and psychical life with the animal creation, but his uniqueness, also contributes to his true essence... The fact that man shares some psychical characteristics with certain primates is totally irrelevant (but true according to Genesis 2) as to his essential nature. Man is God's image and God is spirit. And because man was glorious and his nature so blessed, his fall [into sin] was so tragic.[12]

If man was created in the image of God, and we believe he was (Gen. 1:27), in what ways is man separate from the animal kingdom?

1. Only man has a will and can act on it.
2. Only man has the capacity of the "word" (language/speech).
3. Only man has true self-consciousness.
4. Only humans have the power of decision-making.
5. Man has the power of procreation.
6. Humanity's unique ethico-religious characteristics.
7. Man alone is a moral being.[13]

It is strange that Dr. Collins, who is a born-again Christian, never mentions any of the above unique characteristics of man in his book, *The Language of God: A Scientist Presents Evidence for Belief.* Why is this?

But there is further evidence for the separateness between man and the animals. Even though the animals and man were created on the same day, Day 6, they were created separately and in two different ways. Regarding the land animals, Gen. 1:24-25 says, "And God said, 'let the land produce living creatures according to their kinds... and it was so. God made the wild animals according to their kinds...'"

Regarding man, Gen. 1:26-27 and 2:7, 21-22 records,

> *Then* God said "Let us make man in our image, in our likeness, and let them rule over the fish of the sea and the birds of the air, over the livestock, over all the earth, and over all the creatures that move along the ground. So God created man in his own image, in the image of God he created him... The Lord God formed the man from the dust of the ground and breathed into his nostrils the breath of life, and the man became a living being... So the Lord God caused the man to fall into a deep sleep; and while he was sleeping, he took one of the man's ribs

and closed up the place with flesh. *Then* the Lord God made a woman from the rib he had taken out of the man, and he brought her to the man. (emphases added)

The above *then* is an adverb denoting sequence; first the animals were created, then man and the woman in two separate creative acts. Another indication of man's separateness from the animals is the mandate he received from His Creator to rule over all the animals and over all the earth. Also, the apostle Paul, reflecting on the creation of living creatures, confirms that essential difference in nature, when he says, "All flesh is not the same: Men have one kind of flesh, animals have another, birds another, and fish another" (1 Cor. 15:39).

A second objective that Francis Collins has in placing so much emphasis on the mapping of the human genome is this:

2. He wants to convince us that modern humans are descended from the chimpanzee through a common ancestor.
This is what he writes:

> The placement of humans in the evolutionary tree of life is only further strengthened by comparison with the closest living relative, the chimpanzee. The chimpanzee genome sequence has now been unveiled, and it reveals that humans and chimps are 96 percent identical at the DNA level (pg. 137).

Collins tries to bolster this presumptive comparison by the following assumption of population geneticists.

> Population geneticists, whose discipline involves the use of mathematical tools to *reconstruct* the history of populations of animals, plants, or bacteria, look at these facts about the human genome and conclude that they point to all members of our species having founders, *approximately* 10,000 in number, who lived *about* 100,000 to 150,000 years ago. This information fits well with the fossil record, which in turn places the location of those founding ancestors *most likely* in East Africa.[14] (emphasis added; see Figure 2.1)

Notice the uncertainties in the above statement: "reconstruct;" "approximately;" "about;" and "most likely." There is nothing empirically certain about these assumptions.

Biochemist Fazale Rana questions the ninety-six percent figure, "but asserts that in any case common sense tells us these types of genetic comparisons are meaningless because they do not explain the biological and behavioral differences between chimps and humans. Rana also says close genetic similarity does not require shared ancestry."[15]

What is more, present-day speculation about human evolution revolves around a group of fossils in East Africa called *Australopithecus* and, in particular, a specimen called Lucy; a 40% complete skull. Other skulls have been discovered as well. At best, these bits and pieces of skulls, bones and teeth are extremely flimsy evidence of the so-called "founders" of the human race that Collins cites above in support of man's descent from a common ancestor. Neither anthropologists, nor paleontologists, nor evolutionists have been able to discover one missing link between the ape and man.

Furthermore, there is not even a hint in Genesis 1 and 2 or the rest of the Scriptures that this pre-Adamic man or population ever existed. And why should it appear in the sacred record? That was never in God's plan when He decided to create the universe, planet Earth, and man to inhabit it, and to glorify Him and to enjoy Him forever.

Dr. Collins' third and final objective (reason) for his speaking so highly of the human genome discovery is,

3. To prove that Charles Darwin was right in postulating his theory of biological evolution and that the Human Genome Project now proves his theory correct.

After explaining how the human DNA matches up so closely with the DNA of other non-human primates, Dr. Collins asks, "What does all this mean? At two different levels, it provides powerful support for Darwin's theory of evolution, that is, descent from a common ancestor with natural selection operating on randomly occurring variations" (ppg. 127, 129).

In another place in the same chapter, he writes, "The study of genomes leads inexorably to the conclusion that we humans share a common ancestor with other living things" (pg. 134).

Collins says that

in the nineteenth century, Darwin had no way of knowing what the mechanism of evolution by natural selection might be. We can now see that the variation he postulated is supported by naturally occurring mutations in DNA. These are *estimated* to occur at a rate *of about* one error every 100 million base pairs per generation... Most of these mutations [changes] occur in parts of the genome that are not essential, and therefore they have *little, or no* consequence... But on *rare* occasions, a mutation will arise *by chance* that offers a *slight degree* of selective advantage. That new DNA 'spelling' will have a *slightly higher likelihood* of being passed on to future offspring. *Over the course of a very long period of time,* such favorable *rare events can become widespread* in all members of the species, *ultimately resulting in major changes* in biological function. (pg. 131, emphases added)

Again, notice the terms of plausibility used in the above explanation: "estimated;" "of about;" "little or no;" "by chance;" "slight degree;" "slightly higher likelihood;" "over the course of a very long period of time;" "rare occasions;" "rare events can become widespread;" and "ultimately resulting in major changes." What is the probability of all that coming to pass?

In the above statement about mutations, or changes over time, Collins contends that the mechanism that propels evolution along through a very long period of time is the naturally occurring mutations in DNA. However, we would ask, "Is this scientific fact or fiction?" Has he ever observed these changes taking place?

Francis Collins is a Darwinian Evolutionist, and evolutionists always start with the assumption that man has evolved from the ape or chimp, and then attempt to assemble the evidence to support their claim. This is exactly what Dr. Collins does in this case.

First, he tries to prove descent from the chimp because the chimp's genome displays "near identity" with the human genome. But other scientists do not agree, saying that close genetic similarity does not require shared ancestry. Besides, this genetic similarity does not explain the fundamental biological and behavioral differences between chimps and humans. They were created by God in separate acts and, therefore, are not related genetically.

Second, Collins attempts to prove descent from a common ancestor by the use of mutations in DNA. But these mutations were computer and mathematically generated. They were assumed to have taken place, but no one has actually verified these changes. So this assumption belies the first component of the scientific method, which is: Are they (the mutations) observable? This is the first step in attempting to prove a theory.

Furthermore, Genesis 1 and 2 singularly reflect a mature creation from the beginning. Note these aspects, if you will:

Aspects of a Mature Creation (partial list)

- •Continents with topsoil
- •Plants bearing seed
- •Fruit trees bearing fruit
- •Land with drainage system
- •Rocks with crystalline minerals
- •Rocks with various isotopes
- •Stars visible from earth
- •Marine animals adapted to ocean life
- •Birds able to fly
- •Land animals adapted to environment
- •Plants and animals in symbiotic relationships
- •Adam and Eve as adults
- •All "very good"[16]

Figure 5.1

If the various aspects of creation including the man and woman were mature from the beginning of time and were done on six separate days of twenty-four hours each, then there was no need for mutations over extended periods of time. The Genesis account is very clear that Adam and Eve were created adults, and being created in the image of God, that they possessed both the natural and moral images of God at creation. We can only conclude, then, that the "descent of man by change" and its foundational assumption of evolution are scientific fiction, with no basis in fact.

Concluding Remarks

Dr. Collins and his team of researchers should be applauded for the marvelous discovery of the human genome. There is no doubt but that this discovery will be very beneficial to humanity in helping researchers and medical professionals to better understand human diseases and to treat them more effectively. Humanity is afflicted with so many incurable diseases that it would be advisable for medical science to expend its time and energy in this realm alone (the scientific) without making forays into the spiritual realm, by way of biological assumptions and pronouncements. The spiritual is the province of God and Christian theology alone. God has already spoken in His Word about how he created man, who man is, why he is here, and where he is going when he dies. And this cannot be infringed upon by science. In chapter six we will deal with Francis Collins' theory of *Theistic Evolution.*

Endnotes

1. Richard Ostling, "The Search for the Historical Adam," *Christianity Today*, June 2011, 23.

2. Loc. cit.

3. *Webster's Ninth New Collegiate Dictionary*, (Springfield, MA: Merriam Webster Inc. Publishers, 1986), 1009.

4. Loc. cit.

5. Ostling, op. cit., 27.

6. Tim Dowley, ed., *A Lion Handbook: The History of Christianity* (Herts, England: Lion Publishing, 1977), 490.

7. *The Hymnal for Worship and Celebration* (Irving, TX: Word Music, 1989), 273.

8. ESV Study Bible (Wheaton, IL: Crossway, 2011), 977.

9. C. Wesley King, *Creation for Earnest Believers* (Nicholasville, KY: Schmul Publishing Company, 2012), 103-107.

10. For further elaboration of these two points, see my book *Creation for Earnest Believers,* 106-107.

11. Ibid, 104.

12. Ibid, 108.

13. Ibid, 108-111.

14. Francis S. Collins, *The Language of God* (New York: Free Press, 2006), 1216.

15. Ostling, op. cit. 25.

16. John D. Morris, *The Young Earth* (Green Forest, AR: Master Books, 1997), 39.

6

A Biblical Response to and Critique of Francis Collins' Book, The Language of God *(continued)*

Chapter Ten, "Option 4: BioLogos (Science and Faith in Harmony)"

Introduction

PART I OF THIS BOOK is titled "Theistic Evolution— The Liberal View (From Charles Darwin to Francis Collins)."

In chapter four, "A Timeline of Growing Evangelical Acceptance of Theistic Evolution," I referred to Carl F. H. Henry's thought-provoking book, *Remaking the Modern Mind,* whose message is even more relevant today than when it was written in 1946. Writing at the close of World War II's devastation, he and others were calling for a remaking of the modern philosophical mind that was predicated on the premises that nature is the ultimate reality and that man is only an animal (see psychologist B. F. Skinner, for example). This has morphed into a belief that man is a "special class of animal" (Collins, *et al.).* Henry correctly contends that over the last 350 years the modern mind has been shaped by rationalism, empiricism, and positivism that left no room for supernatural revelation and the belief that man stands in unique relation to his Creator and to the objective, eternal moral order. Henry goes on to assert that *naturalism* (the belief that "nature is all there is") made the break with the traditional, biblical view of God, man and the universe of the pre-modern era complete, discarding the notion of a projected Absolute as vigorously as it discarded the idea of a divine revelation. With Hume, Comte and finally Dewey, this naturalistic philosophy captured the western mind. "A century ago [1859] modern philosophy took for its bride the scientific theory of evolution; in the course of the long wedlock there was occasion for many *off-*

spring... "[1] (emphasis added). Among the *offspring* generated by Darwin's theory of biological evolution is *Theistic Evolution* which Francis Collins accepts and promotes wholeheartedly.

How Collins Came to Accept Theistic Evolution

In the introduction, Collins highlights the central question of his book as being this:

> In this modern era of cosmology, evolution, and the human genome, is there still the possibility of a richly satisfying harmony between the scientific and spiritual worldviews? I answer with a resounding yes! In my view, there is no conflict in being a rigorous scientist and a person who believes in a God who takes a personal interest in each one of us. (pg. 6)

Later in this chapter he says,

> I confess that I didn't pay much attention to the potential for conflict between science and faith for several years— it just didn't seem that important. There was too much to discover in scientific research about human genetics, and too much to discover about the nature of God from reading and discussing faith with other believers.[2]

Then in his own words he says,

> The need to find my own harmony of the worldviews ultimately came as the study of genomes— our own and that of many other organisms on the planet— began to take off, providing an incredibly rich and detailed view of how *descent by modification from a common ancestor* occurred. Rather than finding this unsettling, I found this elegant evidence of the *relatedness of all living things* an occasion of awe, and came to see this as *the master plan* of the same Almighty who caused the universe to come into being and *set its physical parameters just precisely right to allow the creation of the stars, planets, heavy elements, and life itself.* Without knowing its name at the time, I settled comfortably into a synthesis generally referred to as "theistic evolution," a position I find enormously satisfying to this day. (ppg. 198, 199, emphasis added)

What is Theistic Evolution?

A concise definition of Theistic Evolution is this: Theistic Evolution is the attempted harmonization of evolution with a belief in God. More particularly, it is a belief in God, and that God used natural processes as His method of creation, guiding evolution to the final realization of man.

In this connection, Collins admits that there are many subtle variants of Theistic Evolution, but a typical version rests upon the following six premises, which he presents as his own version (pg. 200):

1. The universe came into being out of nothingness, approximately 14 billion years ago.
2. Despite massive improbabilities, the properties of the universe [molecules, atoms and chemical elements] appear to have been precisely tuned for life.
3. While the precise mechanism of the origin of life on earth remains unknown, once life arose, the process of evolution and natural selection permitted the development of biological diversity and complexity over very long periods of time.
4. Once evolution got under way, no special supernatural intervention was required.
5. Humans are part of this process, sharing a common ancestry with the great apes.
6. But humans are also unique in ways that defy evolutionary explanation and point to our spiritual nature. This includes the existence of the Moral Law (the knowledge of right and wrong) and the search for God that characterizes all human cultures throughout history.

A little later on, I will point out the flaws and fanciful assumptions in the above premises. After stating these six premises as his belief system, Collins goes on to say,

> If one accepts these six premises, then an entirely plausible, intellectually satisfying, and logically consistent synthesis emerges. God, who is not limited in space or time, created the universe and established laws that govern it. Seeking to populate this otherwise sterile universe with living creatures, God chose the elegant mechanism of evolution to create microbes, plants, and animals of all sorts. Most remarkably, God intentionally chose the same mechanism to give rise to special creatures who would have intelligence, a knowledge of right and wrong, free will, and a desire to seek fellowship with Him. He also knew these creatures would ultimately choose to disobey the Moral Law.
>
> This view is entirely compatible with everything that science teaches

us about the natural world. It is also entirely compatible with the great monotheistic religions of the world [Judaism, Christianity, and Islam]. The theistic evolution perspective cannot, of course, prove that God is real, as no logical argument can fully achieve that. [That God exists and is real is revealed truth in the Christian Scriptures and is experienced by faith.] But this synthesis has provided for legions of scientist-believers a satisfying, consistent, enriching perspective that allows both the scientific and spiritual worldviews to coexist happily within us. This perspective makes it possible for the scientist-believer to be intellectually fulfilled [first] and spiritually alive [second], both worshipping God using the tools of science to uncover some of the awesome mysteries of His creation. (ppg. 200, 201)

This is the essence of what Theistic Evolutionist Francis Collins believes about the universe, Planet Earth, and the process that brought living things including man into existence.

In the following chart is my rebuttal to Collins' six premises. My remarks are based on Genesis 1 and 2, Psalm 33:6-9 and John 1:1-3.

Theistic Evolution	**Biblical Creationism**
Based on human premises and assumptions	Based on divinely-revealed truth
Age of the created universe	*Age of the created universe*
1. The universe came into being out of nothingness, approximately 14 billion years ago [as a result of the "Big Bang"]. (Old Earth)	1. 6,000 to 12,000 years ago out of nothing *(ex nihilo)* (Young Earth)
2. Despite massive improbabilities, the properties of the universe [chemical elements, molecules and atoms] appear to have been precisely tuned for life.	2. The properties (chemical elements, molecules and atoms) that make up matter and the universe were brought into existence by the creative action of Almighty God when he created *(bara)* the heavens and the earth "in the beginning." This verb denotes that which is

Theistic Evolution

Biblical Creationism

brought about by the activity of divinity— by God only.

When God creates *(bara)* something, He does that which has no human or self-originating parallel. His word or creative power effects His will, even creating the necessary material of which an object is to consist.

3. While the precise mechanism of the origin of life on earth remains unknown, once life arose, the process of evolution and natural selection permitted the development of biological diversity and complexity over very long periods of time.

3. The origin of life on earth is known for it is clearly revealed in God's Word. On the fifth day God created *(bara)* the lower animals, the birds of the air, and the fish of the sea— an entirely new form of life (1:21).

On the sixth day, God created *(bara)* man in his own image (1:27).

God leaves no doubt as to who was the Creator. The animal world and man did not come into existence through some mechanical principle or process of nature at work in the lower forms of life. They came into existence because they were called into existence by God Himself.

Biological diversity and complexity in plant, animal and human life is attributable to God's will

90   THE BATTLE FOR GENESIS 1 AND 2

Theistic Evolution	Biblical Creationism
	and design for each of these forms of life when He brought them into existence.
	Naturalistic evolution breaks down at three vital points:
	(1) it has not been able to bridge the chasm between the inanimate and the animate;
	(2) it cannot pass from the diffused life of vegetable realm, to the conscious somatic life of the animal kingdom; and
	(3) it cannot pass from the irrational life of animals to the rational self-conscious life of man.
	Only the creative activity of God could have originated vegetable, animal and personal life.
	The theory of the differentiation of species breaks down further in the case of the sterility of hybrids.
	The declaration in the Genesis account that each shall bring forth after its kind is an acknowledged fact, both in the realm of science and in the world of experience.
4. Once evolution got under way, no special supernatural intervention was required.	4. Evolution over long periods of time involving the material and human worlds was unnecessary and never existed

Theistic Evolution	Biblical Creationism
	in God's mind or will because God called everything He created into existence by command, or *fiat.* For example, Psalm 33:9 says, "For he spoke, and it came to be; he commanded and it stood firm."

The creation of the universe, Planet Earth, living things and man was supernatural rather than natural.

This is borne out further by the following: In biblical thought God's word (*däbär*) and God's deed (*däbär!*) are one and the same. Especially is this evident with respect to God, for His word is His creative word. To do was to speak, to speak was to do. God's word resulted in action, namely, the immediate, yet successive, creation of the world and all things in it.

The Genesis 1 text says, "And God said, 'Let there be... and it was so'" (1:3, 6, 9, 11, 14, 20, 24, and 26). Just as soon as God spoke, all of the things that He was calling into existence were created immediately— light, the firmament, dry land, plants and trees, the sun, moon and stars, fish and

Theistic Evolution

Biblical Creationism

fowl, land animals and man.
In the case of plants and trees,
fish and fowl, animals and
man, these were all created
mature with the capacity to
begin reproducing immedi-
ately according to their kinds
or species (1:11, 12, 21, 24
and 25).

5. Humans are part of this process [evolution], sharing a
common ancestor with the
great apes.

5. Human beings are *not* and *cannot* be a part of an evolutionary process, nor do they share
a common ancestor with the
great ape for the following
reasons:

(1) There is a distinct break in the
sacred record between the
creation of the land animals
(Gen. 1:24-25: "Let the land
produce living creatures according to their kinds... ") and
the creation of man and
woman (Gen. 1:26-27: *"Then*
God said 'Let us make man
in our image...' So God created man in his own image, in
the image of God he created
him; male and female he created them.")

(2) There is no mention or hint
whatsoever of evolution or of
man being a part of any evolutionary process in Genesis
1 and 2.

Theistic Evolution	Biblical Creationism

Biblical Creationism

(3) Man was definitely created in the natural and moral image of God himself, whereas the animals were created with neither.

(4) As pointed out earlier, the verb *bara* "to create" denotes that which is brought about by the activity of divinity. Man was supernaturally created by God— a special creation— as opposed to being the result of a natural process.

Theistic Evolution

6. But humans are also unique in ways that defy evolutionary explanation and point to our spiritual nature. This includes the existence of the Moral Law (the knowledge of right and wrong) and the search for God that characterizes all human cultures throughout history.

Biblical Creationism

6. While it is true that man shares on an elementary level physical and psychical life with the animal kingdom, there is abundant experiential knowledge and evidence to prove that man is not merely an extension of primates or that man and the great ape have a common ancestor, but that man is uniquely different from the animal kingdom and that he exists on a separate level far above the animals, insomuch as he was created *(bara)* in the image of God (1:27) and by the hand of God (2:7)— a special creation (Psalm 8:4-9).

This sacred truth has been held by the Jewish and Christian

Theistic Evolution	Biblical Creationism
	communities from the beginning of time.
	Man's uniqueness consists in his having been created in the image of God, which includes his natural image (understanding, freedom of will and affections)— marred by sin in the Fall, and his moral image (righteousness and true holiness)— lost because of sin.
	The image of God in man also includes man's power of thought, his power of communication, his power of self-transcendence, and his consciousness or moral responsibility. This he received directly from his Maker.
	Man's uniqueness contributes to his true essence. The fact man shares some physical characteristics with certain primates is totally irrelevant (but true according to Genesis 2) as to his essential nature.

An Evaluation of Collins and His Beliefs

Agreement with Creationists. He has two things in common with Creationists. First, he believes that the universe came into being out of nothingness. Second, he believes that God created matter, the universe and the laws that govern it. This seems to be the extent of his theism and accounts for his use of "theistic" with "evolution."

Otherwise, Collins is essentially a Darwinian Evolutionist who believes:

1. In a very old earth and universe.
2. In man's descent from a common primate ancestor with natural selection operating on randomly occurring variations.
3. In mutations (changes) in the DNA of the genome arising by chance and occurring over the course of a very long period of time. These changes offer a slight degree of selective advantage. Such favorable rare events can become widespread in all members of the species, ultimately resulting in major changes in biological function.
4. In the relatedness of humans with all other living things.
5. That God chose the mechanism of evolution (how does he know this?) to create microbes, plants and animals of all sorts.
6. That God intentionally chose the same mechanism to give rise to special creatures (human beings) who would have intelligence, a knowledge of right and wrong, free will, and a desire to seek fellowship with Him. He also knew these creatures would ultimately choose to disobey the Moral Law.

Collins' views are heavily weighted on the side of evolution. So it would be more accurate and honest to describe the two worldviews he calls scientific and spiritual as naturalistic and theistic.

He defends Darwinian Evolution even as he insists on God as the Creator, but by not upholding Biblical Creationism as recorded in Genesis 1 and 2, he is not only at the epicenter of a dispute with fellow believers, but also doing great harm to the evangelical cause.

He is straddling the fence between true theism and Darwinian Evolution, as all Theistic Evolutionists do, by believing in God but assigning the creation of all living things to evolution. Thus he denies God's supernatural role and activity throughout the entire creative process.

Problems Inherent in Collins' Premises

Premise 1. It is good that Collins recognizes that the universe came into being out of nothingness (Heb. 11:3) but what is the scientific evidence that it came into being approximately 14 billion years ago? "Approximately" is not exactitude or empirical evidence.

Premise 2. "Despite massive improbabilities, the properties of the universe appear to have been precisely tuned for life." What might have been the statistical probability that the universe was "precisely tuned for life"? Who did this tuning? How was it done? Did it happen on its own— that is, was it self-generated by nature— or did Someone do it?

Premise 3. "While the precise mechanism of the origin of life on earth remains unknown, once life arose, the process of evolution and natural selection permitted the development of biological diversity and complexity over very long period of time." Notice the inconsistency and illogic in such a statement.

• We don't know the *precise mechanism* of the origin of life on earth, assuming it was a "mechanism" at all.

• We don't know *when* life arose.

• We don't know *what* it was about the process of evolution in itself that permitted the development of biological diversity and complexity over very long periods of time.

• We don't know how long it took for evolution to develop one biological form into another one, or why it took so long. Such a nebulous process doesn't inspire confidence to believe it.

These are very serious flaws in Collins' third premise. Let's look at it this way. In his book, *Studies in the Bible and Science,* Henry Morris explains logically what Collins can only assume in Premise 3.

As to who or what began it [the universe], we can conceive of only two alternatives. There is the possibility that some impersonal principle (call it chance or what you will), something intrinsic in the natural order of things, about whose origin we can know nothing, somehow shaped things into their initial form, and then set them to following out a deterministic process of development, or rather degeneration. Or, there is the possibility that all things were created in the beginning by a Person, also about whose origin we know nothing. It comes to this: The universe was begun either by a Person, or by something without personality.

But if the law of cause and effect means anything, the universe could only have been brought into existence by a *cause* adequate to account for everything, every concept, every character exhibited by the universe. It is axiomatic, at least as far as anything now going on in the universe is concerned, that the effect cannot be greater than the cause. A cause must have at least all the characters of the effect it produces. How then, can it be possible, even by a nearly interminable process of evolution to produce intelligence, to produce feeling, emotion, will— in short, to produce personality (which is beyond doubt an effect observable in the universe and in our own self-consciousness), if the cause is not itself possessed of personality?[3]

Premise 4. Once evolution got under way, no special supernatural intervention was required.

This statement is antithetical to the above philosophical assumption of Morris. If we believe that evolution does not have personality, then it cannot be the cause of everything in the universe. Premise 4 relegates God to the sidelines— God is "benched" in favor of a "better player." Evolution trumps God in the ongoing creative process. To say that no special supernatural intervention is required is to deny that God is the real Creator of everything in the universe.

Premise 5. "Humans are part of this [evolutionary] process, sharing a common ancestor with the great ape." Webster defines evolution as "a process of continuous change from a lower, simpler, or worse to a higher, more complex, or better state."[4] Continuity in nature is the essence of modern evolutionary thought. Evolutionists claim that the mechanism for evolution is natural selection, chance, and long periods of time. Premises 3, 4 and 5 assume a process of evolution or continuity and are central to Collins' theory of Theistic Evolution. But is he correct in assuming that humans are part of this [evolutionary] process?

As noted above in my rebuttal of Premise 3,

if the law of cause and effect means anything, the universe could only have been brought into existence by a *cause* adequate to account for everything, every concept, every character exhibited by the universe...

Therefore, the only reasonable [and logical] conclusion, if causality means anything, is that the universe and everything in it was created by a great personality— by God.[5]

According to conservative theologian Carl F. H. Henry, "the basic problem here concerns the cause-effect relationship. The prime assumption of modern evolutionary thought is that causality flows uphill."[6] He continues to point out that evolution offers no example that materiality arises from non-material (space-time), or life from the non-vital, mind from a non-mental matrix, or consciousness from non-consciousness.[7] The paradox of the universal acceptance of evolution is that it exists despite the basic deficiencies in its evidential basis.

The only logical conclusion, then, is the belief in the special creation of life, of the major forms or organisms, and of man, just as the Lord God revealed it, and as it was recorded by Moses in Genesis 1 and 2.

Premise 6. This premise fails to have any merit for the reasons state above and because man was a special creation by a loving, all wise, and sovereign God, with whom He desired to have daily communion (Gen. 3:8b).

By way of summary, I wish to contrast the views of Francis Collins with those of a true Creationist.

A Theistic Evolutionist versus a True Creationist

Francis Collins	*A Genesis Believer*
1. Denies the validity of Genesis 1 and 2 as interpreted by Young Earth Creationists (YEC) (pg. 172).	1. Accepts the literal rendering of Genesis 1 and 2 as divinely-revealed truth.
2. Accepts macroevolution, the belief that allows one species to evolve into another (pg. 172).	2. Generally accepts the idea of microevolution, the possibility that small changes may occur within the kinds originally created by God, but these variations cannot step over prescribed boundaries.

Francis Collins	*A Genesis Believer*
3. Believes that living creatures came into being by an evolutionary process chosen by God (pg. 201).	3. Believes that all living creatures came into being by creative acts of God called *fiats*.
4. Holds to the biological relatedness of man to all other living things (pg. 141).	4. Holds to the distinctiveness of man among all living things as a special creation of God despite the near identity of the human genome with that of the chimp.
5. Denies the supernatural intervention in the creative process except for the universe (pg. 200).	5. Accepts the supernatural action of God in the creative process from beginning to end, Genesis 1:1-2:25.
6. Believes that dinosaurs lived and died before humans appeared on the earth (pg. 95).	6. Believes that all land animals, including the dinosaurs, and man were created on the sixth day in separate acts by God, and that they lived at the same time.
7. Neither Adam and Eve, nor Job and Jonah were historical figures because there were no eyewitnesses to their existence (pg. 209).	7. Believes that all four individuals were historical figures because they are referred to as such by Jesus and other Old and New Testament writers in numerous passages.
8. Believes that man and the chimp share common ancestry with a prior primate species. The near identity genetic match of ninety-six percent between the two is almost certain proof of this descent (pg. 127ff).	8. Believes that the chimp species was created by God as were humans. Genetic comparisons are meaningless because they do not explain the fundamental biological and behavioral differences between chimps and humans.

Francis Collins	*A Genesis Believer*
9. Believes that all members of our species are descended from a common set of founders, approximately 10,000 in number, who lived about 100,000 to 150,000 years ago (pg. 126).	9. Creationists believe that the story of a pre-Adamic population on earth is a false narrative and specious.
10. Adam and Eve are a symbolic allegory of the entrance of the human soul into a previously soulless animal (pg. 206-207).	10. Adam and Eve were created perfect individuals with both a body and a spirit or soul. Together they comprise human personality. They were created in both the natural and moral images of God. The natural image consists of immortality, understanding, free will and emotions. The moral image of God, according to the Apostle Paul, is "righteousness and true holiness" (Eph. 4:24). This was lost in the Fall.
11. According to Francis Collins man is a "special class of animal"— presumably created in the "image of the great ape" (pg. 133).	11. According to the Word of God, man as the crown of God's creation was created in the very image of God Himself. What a magnificent truth! (Psalm 8)

Final Thoughts

When a friend of mine asked me two years ago if I had read this book, I said "No, but I thought perhaps I should." So I bought it and read it with interest and dismay, but not with profit. I can honestly say that I hold no animus toward Dr. Collins— only great disappointment with what he has written and the stand that he takes in defending Theistic Evolution, which is indefensible and untenable in the light of God's Word, especially Genesis 1 and 2. In Isaiah 40:6b-8 God declares, "All men are like grass, and all their glory is like the flowers of the field. The grass withers and the flowers fall, because the breath of the Lord blows on them. Surely people are grass. The grass withers and the flowers fall, but the word of our God stands forever." What God has said about man in Genesis 1 and 2 He means, and He has written it in understandable language. It cannot be changed or amended, and it can only be denied at great cost. I am convinced that Dr. Collins wrote this book in 2006 to sell a product— Theistic Evolution, and that he established the BioLogos Foundation in 2007 to promote it, especially among evangelicals.

After carefully studying the book, the Genesis account of creation is still more intellectually satisfying and logically consistent than the Theistic Evolution account. The book is full of "newspeak," or double talk, a language marked by ambiguity and contradictions, and most of all, it is dishonoring to our holy and sovereign God. What Dr. Collins is asking Bible-believing Christians to do is to exchange our "Cadillac plan" of creation— which is working just fine— for a cheaper plan which is based on wobbly premises and unproven assumptions that require more faith to believe than the God-given plan in Genesis.

The Language of God denies many fundamental beliefs of the Christian faith that the Church has held sacred and inviolable for 2000 years.

First, the book denies the possibility of the Genesis days being literal twenty-four-hour days in favor of long periods of time.

Second, the book denies that Adam and Eve were the first parents of the entire human race, claiming there was a sizeable pre-Adamic population.

Third, the book denies that the Adam and Eve of Genesis were a historical couple, a true narrative, but rather a myth and a poetic and powerful allegory of the entrance of the human soul and moral nature into a previously soulless animal kingdom.

Fourth, more seriously, the book claims that evolution, rather than Christ the Word, was the instrumental means whereby God created all things. This borders on blasphemy.

Fifth, the book asserts that human beings along with the great ape share a common ancestry with prior primate species.

Sixth, the book denies God his rightful place as *sovereign and sole Creator* of the universe and everything in it, both seen and unseen.

Seventh, the book denies Christ the Word (John 1:13; Col. 1:15-17) His rightful place as *sole instrumental means* in creation.

Eighth, the book denies the opening chapters of Genesis that are indispensable for the later unfolding of God's plan for redeeming fallen humanity. Both creation and redemption are inextricably bound together and inseparable biblically and philosophically. Both require faith and not scientific evidence.

Ninth, the book will bring further division in the Body of Christ. Even now it is challenging man's unique status as bearing the "image of God." It is on the verge of undermining Christian doctrine on original sin and the Fall, the genealogy of Jesus in the Gospel of Luke, and most significantly, Paul's teaching that links the historical Adam with redemption through Christ.

Our God deserves all the glory that comes as the result of human discoveries of truth, including the human genome. But He has warned in His Word, "I will not give my glory to another" (Isa. 42:8; 48:11).

Endnotes

1. Carl F. H. Henry, *Remaking the Modern Mind* (Grand Rapids: Wm. B. Eerdmans Publishing Company, 1946), 23-24.

2. Francis Collins, *The Language of God* (New York: Free Press, 2006), 198.

3. Henry Morris, *Studies in the Bible and Science* (Philadelphia: Presbyterian and Reformed Publishing Co., 1967), 13-14.

4. *Webster's Ninth New Collegiate Dictionary*, (Springfield, MA: Merriam Webster Inc. Publishers, 1986), 431.

5. Morris, op. cit., 14.

6. Henry, op. cit., 148.

7. Ibid, 143.

PART 2
THE ROLE OF SCIENCE IN RELATION TO
REVEALED TRUTH ABOUT CREATION

7

Science: Origin, Disciplines, the Scientific Method and Departure from Divine Truth

Introduction

WE ARE NOW LIVING IN the post-modern era, when truth is relative; it is whatever the group says it is. In postmodernism the rational is replaced by rhetoric. Words are replaced by images, reason by emotional gratification, morality by relativism, meaning by entertainment, truth by fiction. In today's society it does seem at times that truth has perished from our land and that it has vanished from our lips (Jer. 7:28). But this is only a human and secular perspective on life. The divine reality is that truth still exists— it is that absolute, objective and ultimate truth that comes from above.

The God of the Universe is the Source of all Truth and Knowledge

Psalm 90 is a prayer of Moses, the man of God. In this psalm he establishes the eternalness of the God of the universe and the finiteness of man. In Colossians 2:1-5 the Apostle Paul offers a prayer for the Colossian believers, "that they may have the full riches of complete understanding, in order that they may know the mystery of God, namely, Christ, [the Son of God] in whom are hidden all the treasures of wisdom [Gr., *sophia*] and knowledge [Gr., *gnosis*]. Here Paul emphasizes that Christ [who is the Word spoken of in John 1:1-3, 14] is the ultimate storehouse of divine wisdom and knowledge. Since this is true, all truth and knowledge— whether revelational, historical, theological, metaphysical or scientific— resides in and flows from the eternal God of the universe, who is the all-knowing and all-wise One. Henry Morris confirms the same when he says, "In the final analysis, all truth is one. God did not create one universe of physical reality and another of spiritual reality. The same God created all things, and His Word was given by His Holy Spirit to guide us into *all* truth."[1]

From this truth flow two corollaries: (1) The One who created the universe knows all about its materiality and how it functions, and (2) He allows man, who is made in His image and endowed with creativity, reason and curiosity, to explore, theorize and discover the wonders of nature and the secrets of the universe, and then to create heretofore unknown scientific fields of study, and to place his findings in these new disciplines such as astronomy, geology, cosmology, archaeology, physics, biology and the like. In a moment we shall notice the origin of these physical and living sciences and their pioneers or founders. Right now we need to examine briefly what was taking place in Europe several centuries ago.

The Transition from the Medieval to the Modern World We Live in

The Protestant Reformation (1517-1648) of which we are the spiritual beneficiaries, was preceded in Europe by the Renaissance (1350-1650). The term *renaissance* literally means "rebirth of culture." The Renaissance, which took place in important countries of Europe, marked the transition from the medieval to the modern world.

This new movement took two different directions. The classical Humanism south of the Alps was matched by the religious Humanism of Reuchlin, Colet, and Erasmus north of the Alps in the sixteenth century. The Northern Humanists went back to the Bible in the original, but the Southern Humanists emphasized the study of the classical literature and treasures of the past and languages of Greece and Rome.

The Renaissance opened up new vistas and challenges to discovery. In short, this was an age of physical explorations of the earth: Columbus (1492)– the New World; Vasco de Gama (1498); Magellan (1522)– the first to circumnavigate the globe; and Pedro Alvares Cabral (1500)– Brazil; as well as the scientific discoveries of the universe by Copernicus, Galileo, Kepler, and Newton.

Another profound change that the Renaissance brought was in the cultural reorientation in which man substituted a modern secular individualistic view of life, which now dominates Western civilization in our day, for the medieval religious corporate approach to life. As Earle Cairns observes, "Attention was focused upon the streets of Rome and Athens instead of upon the streets of the New Jerusalem."[2] The medieval theocratic conception of the world, in which God was the measure of all things, gave way to an anthropocentric view of life, in which man became the measure of all things.

Despite this huge cultural transformation, two profound upheavals in Western thought were soon underway. In *religion* Luther, Calvin, Zwingli, and Knox were leading a revolt from the Roman Catholic Church and a return to New Testament teaching about justification by faith alone, salvation by grace alone, and the Bible as the only authority for faith and conduct for the believer.

In *science,* Copernicus, Galileo, Kepler and Newton were replacing a universe centered on the earth with one centered on the sun. In a sense this religious reformation and the scientific revolution went hand in hand.

James Moore, writing a monograph entitled "The Rise of Modern Science," states that there is distinct and plausible evidence that Protestantism gave rise to modern science.

> To begin with, there is evidence of specific Protestant contributions to the 'scientific revolution'... Tycho Brahe and Johann Kepler, the great astronomers, were both devout followers of Luther... David Chytraeus, a Lutheran theologian, dealt with the new star of 1572 in his commentary on Deuteronomy and published a book on the comet of 1577... Johann Fabricius, a Lutheran layman, first observed sunspots and the rotation of the sun... Philip van Lansberghe, a Reformed minister and renowned astronomer, became the keenest supporter of Copernicus in the Netherlands. His fellow-countryman Isaac Beckman, a scientist and strict Calvinist, was an early defender of the atomistic philosophy.[3]

Meanwhile, in England, the Puritan movement produced The Royal Society in 1661, which was a gathering of Christian scientists to discuss scientific disciplines and experiment.[4] Sir Frederick Catherwood concurs with this view that Protestantism gave rise to modern science. In his monograph "The Bible and Society," he writes,

> ... Science itself is based on Christian teaching. It was belief in a God

of order, a God of reason, and a God of unchanging decrees, which led to the development of the scientific method in the 17th century. *When science forsakes this basis it loses its way.* Some people have made a god of it. Many are now rejecting it altogether. Its hope lies in a return to its Christian basis.[5] (emphasis added)

The Gradual Development of Scientific Disciplines

Copernicus, Galileo, Kepler and Newton were all men of profound faith in God and Bible-believing Creationists. Therefore, it was because of their knowledge of the Scriptures, the influence of the Protestant Reformation, their openness to the God of the universe, and their curiosity to discover the secrets of that universe that led them to be the founders of scientific disciplines such as astronomy, cosmology, mathematics, and physics.

The "Heavenly" Sciences

With the invention of the telescope in 1610 by Galileo Galilei, Renaissance men turned their eyes to the heavens to unlock the secrets of the universe. The first to do this was Nicholas Copernicus, considered by some to be the founder of modern science.

Listed below are the names of early pioneering scientists, their dates, the science that began with them, and the contribution(s) to their scientific field.

Astronomy (1550)— This is the science of celestial bodies, their magnitude, motions, and constitution.

- •*Nicholas Copernicus* (1473-1543), a Polish astronomer and a contemporary of the great Protestant Reformer, Martin Luther (1483-1546); in his work *De revolutionibus orbium coelestium* (1543), laid the foundation for all modern astronomy, while setting forth a bold new theory— that the universe is heliocentric, rather than geocentric.[6]

Copernicus was followed by others in this field of endeavor.

- •*Galileo Galilei* (1564-1643)— an Italian astronomer, physicist, and mathematician; invented the hydraulic balance and discovered the laws of dynamics. His discovery of the four satellites of Jupiter (1610) by the aid of his newly invented telescope revolutionized the study of astronomy.[7]

- *Johann Kepler* (1571-1630), a German astronomer and devout follower of Luther; discovered the three laws of planetary motion.
- *Sir William Herschel* (1738-1827) was an English astronomer whose specialty was galactic astronomy. With this, astronomy leaves our solar system and reaches out into the expanses of the universe. A galaxy such as the Milky Way contains thousands of stars, nebulae and star clusters.

Cosmology (1656)— From the Greek word *kosmos* (see John 3:16.) This is a branch of Christian theology that deals with the origin, purpose and nature of man and the universe. It is also a branch of astronomy that deals with the origin, structure, and space-time relationships of the universe. The pioneering scientist in this field is:

- *Isaac Newton* (1642-1729), an English physicist, mathematician and natural philosopher, while still a young man, became interested in the question of whether there were basic principles that operated throughout the universe. In 1687 he wrote *Principia Mathematica* in which he formulated the law of gravitation. He also discovered differential calculus, and correctly analyzed white light. An interesting insight into his life is that he believed his scientific discoveries were communicated to him by the Holy Spirit, and regarded the understanding of Scripture as more important than his scientific work.[8] Earle Cairns observes that "gravitation provided the key to unify the phenomena of science, until it was replaced by Darwin's concept of growth (1859, 1871), the principle of natural law was considered to be basic, and men came to look upon the universe as a machine or mechanism that operated by inflexible natural laws."[9]

The Scientific Method

With these scientific developments underway, it was only a question of time that someone would come up with a method for carrying on scientific study with a view to proving scientific theories. That person was Francis Bacon (1561-1626), English philosopher and lawyer, who published his *Novum Organum* in 1620. In this work he developed an inductive method of interpreting nature. Using the inductive method, which is also known as the scientific method, the scientist should accept nothing on the basis of authority alone, whether it be that of the Bible, theology, or Church teaching and tradition. The scientist should develop a hypothesis or theory, observe facts concerning his tentative data, check the facts by repeated experimentation, and only then, develop a general law.

This is how the scientific method is supposed to work. The three key and indispensable components of the scientific method are: 1) Are the facts of the hypothesis observable? 2) Are they repeatable? and 3) Are they testable for reliability? All three elements must be present in judging the validity of a theory.

In the case of the age of the universe, for example, some scientists theorize that the universe came into existence 14 billion years ago. But is this really true? Scott Huse asserts, "It is impossible to prove scientifically any theory of origins. This is because the very essence of the scientific method is based on *observation* and *experimentation,* and it is impossible to make observations or conduct experiments on the origin of the universe."[10] The scientific method, therefore, is totally inadequate for proving scientifically any theory of origins.

The same is true of Charles Darwin's theory of descent by change and Francis Collins' Theistic Evolution, both of which are theoretically dependent on natural selection and mutations over eons of time for their realization. But we must ask: Have evolutionists ever observed these changes taking place? And the obvious answer is, no, they haven't observed them. They have only assumed that they occurred. So, once again the scientific method cannot prove their theory true. The theory of evolution, as well as the theory that the universe came into existence billions of years ago, are based on assumptions and are not scientific facts.

Further Development of Scientific Disciplines

Earlier we spoke of the two "heavenly sciences" that were discovered by men of the Renaissance, which was closely associated with the Protestant Reformation, namely, astronomy and cosmology. Both of these emerging sciences are alluded to in the Scriptures, especially in Job, Joshua, Isaiah, 2 Kings, and Matthew.

Two important insights are worthy of note here. With the exception of Galileo, who was an Italian, all of these early scientists were Christian scientists and they were from countries north of the Alps that were directly impacted by the Protestant Reformation: Poland (Copernicus), Germany (Kepler), and England (Newton and Herschel).

The "Earthly" Sciences

Geology (1735) is the science or study of the history of the earth's formation, especially as recorded in rocks and fossils.

Before the nineteenth century the vast majority of scientists and Christians interpreted earth history in terms of Biblical Creationism and catastrophism (a sudden and worldwide Genesis Flood), and consequently, believed in a relatively young earth and short time scale.

New Ideas About the Earth's Age

However, two exceptions to this universal belief were two English geologists, James Hutton (1726-1797) and Charles Lyell (1797-1875), who challenged the biblical view of Creation and the Genesis flood. Raised a Quaker, Hutton eventually rejected the belief in a literal worldwide flood. He argued that the earth's history could best be explained by examining the earth's layers rather than accepting the validity of questionable Jewish records, meaning the Genesis account.[11]

Hutton proposed that the earth had been molded, not by sudden violent events like the Genesis Flood, but by slow and gradual processes, the same processes that can be observed in the world today. This theory became known as "Uniformitarianism." It implied that the earth has living history and a long one (millions of years). Six thousand years is not long enough for such major evolutionary changes to take place. As Henry Morris observes, what Hutton did not recognize or realize was that true science is necessarily limited to the measurement and study of present phenomena and processes. Science deals with the data and processes of the present time, which can be experimentally measured and observationally verified— two of the three key elements of the scientific method. Morris also observes,

The principle of uniformity is a philosophy, or faith, by which it is hoped that these processes of the present can be extrapolated into the distant past and the distant future to explain all that has ever happened and to predict all that will ever happen. But, when viewed in these terms, it is obvious that uniformity is not proved, and therefore is not properly included in the definition of science.[12]

Charles Lyell, who is remembered for his contributions to the development of Uniformitarian geology, was born the year James Hutton died. He was a lawyer, politician, geologist and the author of *The Principles of Geology* (1830-1833). In this work he continues to cast doubt on the truthfulness of the Bible and to reveal his keen interest in evolutionary geology, meaning the earth evolved over long periods of time, even millions of years.

The Geologic Column and Timetable of Earth's History

In consequence of Hutton's and Lyell's expansion of the accepted biblical time frame for the age of the earth (approximately 6000 years) scientists began to think in terms of an old earth (millions of years).

Scott Huse points out that

Among the sciences, historical geology is in a most awkward position in that it deals with past events and is thus forced to rely upon assumptions which may or may not be true. Documented history only goes back a few thousand years. The earliest authenticated written records date back to about 3,500 B.C. Prior to the existence of eyewitnesses, no one can be absolutely certain of what actually transpired. Consequently, there is no direct irrefutable proof as to the process(es) which formed the rocks of the geologic column or their age. Any such determination can only be indirect, based upon assumptions which may or may not be true.[13]

Nevertheless, based upon the assumptions that Uniformitarianism and organic evolution were established scientific facts, geologists during the nineteenth century, began to compile the geologic column (see Figure 7.1). They arranged the earth's strata according to the various types of fossils they contained, especially their *index fossils* (usually marine invertebrates which are easily recognized, assumed to have been widespread in occurrence, and of limited chronological duration thus marking a specific age determination for a rock formation (not pictured here). Strata with simpler fossils (*presumed to have evolved*

CENOZOIC ERA *(Age of Recent Life)*	Quaternary Period
	Tertiary Period
MESOZOIC ERA *(Age of Medieval Life)* RECOMMENDED: (Age of Middle Life)	Cretaceous Period
	Jurassic Period
	Triassic Period
PALEOZOIC ERA *(Age of Ancient Life)*	Permian Period
	Pennsylvanian Period
	Mississippian Period
	Devonian Period
	Silurian Period
	Ordovician Period
	Cambrian Period
PRECAMBRIAN ERA	

Figure 7.1

first) were put on the bottom of the column while strata containing more complex forms (*presumed to have evolved later*) were placed toward the top of the column. Thus, the entire geologic column, which arose theoretically *ca.* 1909, was a fact founded and built on the assumption that organic evolution was a fact.[14] It is most significant to note that the Geologic Column as shown is Figure 7.1 in its ideal, continuous sequence, does not exist anywhere in the world.

Early Departure from Revealed Truth in Genesis

These new beliefs concerning the age of the earth had serious ramifications. First, by rejecting the belief in a literal worldwide flood (Gen. 7-8) Hutton and Lyell undermined the authority of Scripture and cast doubt on the truthfulness of the Genesis account. Second, by expanding the time frame for the history of the earth to millions of years, they provided a grand opening for biblical scholars to come up with dubious interpretations of what is said in Genesis 1.

The Gap Theory. One questionable way of interpreting Genesis 1:2 is the Gap Theory, also known as the Ruin-Reconstruction Theory. Scott Huse describes this theory like this:

> Proponents of this scheme maintain that the proper translation of Genesis 1:2 should actually be rendered: "Now the earth became form-less and empty." The implication is that the original perfect creation came to a sudden and terrible cataclysmic ruin (usually associated with the fall of Lucifer). The earth is considered to be ancient with much of the geologic ages transpiring between Genesis 1:1 and 1:2 in an imaginary gap prior to the six days of *re-creation.* Like the Day-Age Theory, the Gap Theory endeavors to reconcile Biblical creation-ism to the evolutionary framework. It, too, is plagued with many scien-tific and Biblical contradictions and blunders and is not the preferred interpretation.[15]

The Day-Age Theory. This theory proposes that the six days of creation represent periods of time, ages, not literal twenty-four-hour days. The theory is designed to accommodate the geologic ages and is compatible with evolutionary thinking. Again Huse argues force-fully,

> It has been shown to be an unworkable compromise, both Biblically and scientifically... For example, if the days are actually ages, how did

the fruit trees created on the third day survive for ages before the sun was created on the fourth day? Similarly, this theory fails to accommodate the vital symbiotic inter-relationships among plants (third day), birds (fifth day), and insects (sixth day).[16]

Furthermore, the *Key Word Study Bible* affirms,

The day-age theory claims that the Hebrew word yôm, translated "day," is used to refer to periods of indefinite length, not necessarily to literal days. While this is a viable meaning of the word (Josh. 13:1; Job 32:7), it is not the common meaning [in the Old Testament]. Furthermore, the meaning of "evening" and "morning" throughout chapter one seems to point to literal days. The same terms occur in Daniel 8:14, where they refer to literal days.[17]

Then there are well-meaning evangelical scholars who, wanting to be faithful to the rendering of literal days in Genesis 1, have come up with fanciful interpretations of the creation days.

The Pictorial Day Theory. This theory claims that the six days mentioned in Genesis 1 are the six days during which God revealed to Moses the events of creation. But the command to rest on the Sabbath (Ex. 20:8-10) was based on the *fact* that God created the world in six days (Ex. 20:11).[18]

The Intermittent Day Theory. According to this theory the days mentioned are literal days, but that they are separated by long periods of time. [This seems like another concession to evolutionary theory.] But the command in Exodus 20:8-11 would lose its significance unless all of God's creative activities took place within a period of six continuous days.[19]

The Literary Framework View. This view seeks to portray the creation week as if it were a workweek, but without concern for temporal sequence. But why hold this theory? Is not this, too, a minor and indirect concession to the evolutionary hypothesis of periods or ages?

The ordinary Bible-believing, born again Christian today who believes in a young earth and a historical Adam and Eve may be left scratching his head and asking, Why all of these divergent interpretations of what seems to be a straightforward and majestic account of the creation of the world and man, of the Fall (Gen. 1-3), and of the Flood (Gen. 6-9) in the Genesis record.

The answer to that question resides in the recesses of the heart and mind of every evangelical scholar and scientist who espouses one of the above evolutionary accommodated interpretations of Genesis 1. The viewpoint that is set forth in this book and held by this author, because it is supported biblically and scientifically, is—

The Literal-Historical Theory. This interpretation regards the six days of creation as literal twenty-four-hour days that followed in immediate succession. The earth is generally believed to be only a few thousand years old (a range of 6,000-12,000 years); the geologic ages and the concept of organic (naturalistic) evolution are completely rejected. It is the belief that the universe and all things in it, seen and unseen, natural and supernatural, were spoken into existence through the Word— Christ (Heb. 1:3, Psa. 33:6) out of nothing (Heb. 11:3) by miraculous acts (fiats or commands) of God. This belief also includes the creation of all plant and animal life according to its kind, and the special creation of man as the crown of creation, in the image of God (Gen. 1:26-27).

In contrast to the nebulous circumstances surrounding the Big Bang Theory, which seemingly has been proved scientifically that the universe had a "beginning," the Bible clearly states that the universe owes its origin to the omnipotent power and will of the eternal God.

The Invention of the Microscope (ca. 1674)

It was Anton van Leeuwenhoek (1632-1723), a Dutch draper and scientist, and one of the pioneers of microscopy, who in the late seventeenth century became the first man to make and use a real microscope. He saw bacteria for the first time in 1674.

Van Leeuwenhoek made many biological discoveries using his microscope. He was the first to see and describe bacteria, yeast plants, the teeming of life in a drop of water, and the circulation of blood corpuscles in capillaries. During a long life he used his lenses to make pioneer studies on an extraordinary variety of things, both living and non-living, and reported his findings in over a hundred letters to the Royal Society of England (see page 108) and the French Academy.

It was the microscope and the newly invented flask that allowed

Louis Pasteur to disprove a popular belief in the spontaneous generation of life.

Van Leeuwenhoek's work was verified, using the scientific method, and further developed by English scientist Robert Hooke (1635-1703), who published the first work of microscopic studies, *Micrographic*, in 1655. Thus, Robert Hooke's detailed studies furthered study in the field of microbiology in England and advanced biological science as a whole.[20]

Scientific Disciplines (continued)

Entomology (1766) is a branch of zoology that deals with insects. Jean Henri Fabre (1823-1915), a French entomologist, was a pioneer.

Biology (1813)— A branch of knowledge that deals with living organisms and vital processes whether in plant, animal or human life.

As we enter this arena of science, it is best to remember that biology is a relatively new science. When Charles Darwin in 1859 launched his theory of biological evolution, as Phillip Johnson correctly affirms, "scientists of the nineteenth and early twentieth centuries who established Darwinism and materialism as scientific orthodoxy, knew little of biochemistry, and imagined the cell to be something rather simple that could just ooze itself up out of some chemical broth."[21] In fact, Darwin himself, in postulating his theories of mutations and natural selection, assumed that *nature* itself determined the development and progress of all living things.[22]

All of the discoveries and advances of medical science that have been made in the twentieth century occurred during the lifetime of many of us. These discoveries have come to be a modern fulfillment of an ancient insight of the psalmist David who wrote in Psalm 139:13-14, "For you created my inmost being; you knit me together in my mother's womb. I praise you because I am fearfully and wonderfully made; your works are wonderful, I know that full well."

Most of these advances have come in the second half of the twentieth century (1941-2000) in the exceptional and most advanced nation in the history of the world. Take the following advances and procedures, for example: penicillin, blood plasma, antibiotics, drugs, vaccines, disease control/cure, heart transplants and bypass surgery, harvesting and transplanting organs, hip and knee replacement, various kinds of cancer treatment, and more recently, all the research being done in micro- and molecular biology and the complete mapping of the human genome and the discovery of the total genetic heredity encoded in DNA.

Medical researchers have made tremendous strides in conquering diseases and genetic defects, as the following chart reveals, but biological science has a long way to go without entering the spiritual realm and hypothesizing about man's origin.

Cured Diseases in America and the West	Genetic Defects/Disorders- No Cure Yet
Tuberculosis	Cancer/leukemia
Malaria	Parkinson's disease
Small pox	Huntington's disease
Polio	Diabetes
Pneumonia	Down's syndrome
Cholera	Cystic fibrosis
Shingles	Tay-Sachs disease
Mumps	Alzheimer's disease
Measles	Sickle-cell anemia
Chicken pox	Rheumatoid arthritis
Rubella	Rheumatic fever
Rabies	Multiple sclerosis (MS)
Whooping cough	Lou Gehrig's disease (ALS)
Typhoid	Autism
Influenza	Phenyiketomuria (PKU)
	Epilepsy
	Ebola

Anthropology (1822)— the science of man, especially with relation to his physical character, the origin and distribution of races, human environment, social relations, and cultures.

For decades, since Darwin's idea of descent from a common ancestor became well-known in the late 1800s, anthropologists (having to do with man's origin), paleontologists (having to do with fossil remains) and evolutionists alike have been searching for the missing links between apes and man, but these missing links have proved to be very elusive and non-existent.[23]

Paleontology (ca. 1838)— a branch of geology that deals with the study of ancient prehistoric plant and animal life based on fossil remains found in the earth; the study of fossils. John Woodword (1665-1728) was a pioneer in this field. As a science, paleontology never developed until the nineteenth century, probably spurred on by Darwin's writings in 1859 and 1871 and his theory of evolution.

Evolutionists insist that it is the fossil record which provided actual documentation of organic evolution. But Scott Huse contends that "the fossil record reveals an absence of life forms in the lower two-thirds of the earth's crust (the so-called pre-Cambrian period). Then suddenly, in abundant numbers of advanced forms, life appears."[24] Then he adds,

> Many fossils of plants and animals found in the supposed oldest of rocks, when compared with their counterparts, are found to be essentially the same... In short, the fossil record reveals that life appeared abruptly in great diversity and complexity, and abundance without any ancestors from which to evolve! Clearly, this is not evidence for gradual organic evolution. This is supernatural creation and the sudden devastating and catastrophic effect of the Genesis Flood on plant, animal and human life.[25]

Archaeology (1837)— the scientific study of material remains (fossils, relics, artifacts and monuments) of past life and activities, discovered by archaeologists excavating predetermined sites or locations anywhere on earth, but particularly in the Middle East; the remains of the culture of an ancient people.

In the beginning, archaeology arose as an effort to discover previous civilizations and, by chance, made discoveries that substantiated biblical events and places in ancient history: Egypt, Palestine, and Babylonia.

Initially, two important and rewarding discoveries opened up the field of archaeology. The key that unlocked the ancient Egyptian language of *hieroglyphics*— picture writing, a symbol for each word— was the discovery of the Rosetta Stone in 1799 by M. Boussand, a

French scholar, who accompanied Napoleon on his expedition to Egypt. Similarly, the discovery of the Behistun Rock in 1835 by Sir Henry Rawlinson, which was translated fourteen years later, provided the key to the ancient Babylonian *cuneiform* writing, and opened to the world the vast treasures of ancient Babylonian literature.[26]

The Tell-el-Amarna Tablets unearthed in Egypt, dating from about 1400 B.C., and the Code of Hammurabi (a contemporary of Abraham) found in Babylonia, dating from about 2000 B.C., are among the most important archaeological discoveries of the last 150 years.

In the second half of the twentieth century scores of university and seminary Old Testament professors from America and other countries led archaeological expeditions with their students to excavate traditional sites in Palestine (Jerusalem, Jericho, Ai, Lachish, Hebron, Samaria, Megiddo, and Bethlehem). And what they unearthed was a trove of factual evidence that confirmed the truthfulness of Old Testament history as recorded in the Scriptures.

Summary

In this chapter, we have shown that the God of the universe is the source of all truth and knowledge: revelational, historical, theological, metaphysical and scientific. Two corollaries flow from this truth: (1) the One who created the universe and nature knows all about its materiality and how it functions, and (2) he allows man, made in His image and endowed with reason, creativity, and curiosity, to explore, theorize and discover the wonders of nature and the secrets of the universe, and to create scientific fields of study in which to place his findings.

The European transition from a medieval to a modern world gave rise to the Protestant Reformation of Luther, Calvin, Zwingli and Knox, and to the scientific revolution of Copernicus, Galileo, Kepler and Newton. There is enough circumstantial evidence to show that Protestantism gave birth to modern science for these early men of science were also Christians. Isaac Newton believed that his scientific discoveries were communicated to him by the Holy Spirit ("He will lead you into all truth"— John 16:13a), and regarded the understanding of Scripture as more important than his scientific work. This statement reveals the close ties between religion and science in the beginning of the modern era.

This was followed by a brief description of the "heavenly" sciences of astronomy and cosmology and the "earthly" sciences of geology, entomology, biology, anthropology, paleontology, and archaeology, which oftentimes have been appealed to for support for organic evolution.

Then, based on the definition of the scientific method itself, and its three indispensable components— (1) Is the data observable?, (2) Is it repeatable?, and (3) Is it testable?— we have shown that it is impossible to prove any theory of origins of the earth or the descent of man by change from a common ancestor using the scientific method.

Regrettably, the Christian faith and science began to part company when geologists James Hutton and Charles Lyell challenged the time-honored Genesis accounts of Creation and the Flood by rejecting the literal, sudden, and worldwide Flood and by expanding the time frame for the earth's history to millions of years, thus creating the theory of Uniformitarianism.

Because of this departure from the clear literal-historical interpretation of Genesis 1 and 2, other imaginative theories concerning the meaning of "day" in Genesis 1 arose, which were based on long periods of time and evolution as well: The Gap Theory, the Day-Age Theory, the Pictorial Day Theory, the Intermittent Day Theory, and the Literary Framework Theory.

The rise of these alternative theories of the creation days have divided Protestants and evangelicals and allowed them to "pick and choose" what they want to believe regarding revealed truth in Genesis, and perhaps, in other portions of Scripture.

The inventions of the telescope and the microscope were providential, that is, ordered by God to advance science in the modern era and to permit scientists to discover the solar system and beyond, and view the inner world of nature invisible to the naked eye, to the end that men might praise Him for His created works (Psalm 8) and glorify His name.

Even the full mapping of the human genome, revealing DNA, RNA and genes, was permitted now in the twenty-first century by God for research scientists to better understand human diseases and to treat them more effectively, in order to alleviate human suffering. This discovery of the complexity, diversity and beauty of the human cell points to design and an intelligent Designer. The creation of man and woman in the image of their Creator was a special and supernatural creation on the sixth day of creation, just as the Scripture says in Genesis 2:7 and 2:21-22.

In chapter eight, we shall take up the subject of "Science: Definition, Limitations and Agreement with Revealed Truth in Genesis 1-2."

Endnotes

1. Henry Morris, *Studies in the Bible and Science* (Philadelphia: Presbyterian and Reformed Publishing Co. 1967), 120.

2. Earle Cairns, *Christianity Through the Centuries* (Grand Rapids: Zondervan Publishing House, 1958), 284.

3. Tim Dowley, ed., *A Lion Handbook: The History of Christianity* (Herts, England: Lion Publishing, 1977), 42-43.

4. Ibid, 43.

5. David Alexander and Patricia Alexander, eds, Eerdmans' *Handbook to the Bible* (Grand Rapids: William B. Eerdmans Publishing Company, 1976), 60.

6. Marius Forté and Sam Sorbo, *The Answer* (Telemachus Press, 2013), pgs. 201-202).

7. Elizabeth Livingstone, ed., *The Concise Oxford Dictionary of the Christian Church* (Oxford: Oxford University Press, 1987), 205.

8. Dowley, op. cit., 490.

9. Cairns, op. cit., 406-407.

10. Scott Huse, *The Collapse of Evolution* (Grand Rapids: Baker Book House, 1988), 1.

11. Ian Taylor, *In the Minds of Men: Darwin and the New World Order* (Toronto: TYE Publishing, 1984), 29.

12. Morris, op. cit., 152.

13. Huse, op. cit., 9.

14. Ibid, 9-10, 14.

15. Ibid, 32.

16. Loc. cit.

17. *Hebrew Greek Key Word Study Bible*, NIV (Chattanooga, TN: AMG Publishers, 1996), 2.

18. Loc. cit.

19. Loc. cit.

20. www.history-of-the-microscope.org/anton-van-leeuwenhoek.

21. Phillip Johnson, *Defeating Darwinism by Opening Minds* (Downer's Grove, IL: InterVarsity Press, 1997), 75-76.

22. Dowley, op. cit., 6.

23. C. Wesley King, *Creation for Earnest Believers* (Nicholasville, KY: Schumul Publishing Company, 2012), 100.

24. Huse, op. cit., 37.

25. Ibid, 38.

26. King, op. cit., 119-120.

8

Science: Definition, Limitations, and Agreement with Revealed Truth

Introduction

IN CHAPTER SEVEN WE DISCUSSED the origin of modern science, the unfolding disciplines of scientific study, the importance of the scientific method formulated by Francis Bacon in 1620 (the same year the Pilgrims landed at Plymouth in the New World) and the beginning of the departure of early geologists from the traditional and historical understanding of the Genesis account of the creation of the earth and man.

The reader may be wondering why I didn't give a definition of science in chapter seven. The reason is that I wanted to establish in the reader's mind (1) the origin of modern science, (2) its close connection with the Protestant Reformation and the revealed truth of Genesis, (3) the newly discovered disciplines of astronomy, cosmology, geology and anthropology, and (4) a definition of and manner in which the scientific method works as a guide for doing science. With that in mind, let's define science.

Definitions of Science

The word "science" is derived from the Latin *scientia* ("knowledge"), and this is essentially what it means.

Webster defines science as "knowledge covering general truths or the operation of general laws especially as *obtained* and *tested* through the scientific method."[1]

A more formal definition, as given in the Oxford Dictionary is:

A branch of study which is concerned either with a connected body of demonstrated truth or with observed facts systematically classified and more or less colligated by being brought under general laws, and which includes trustworthy methods for the discovery of new truth within its own domain.[2]

Regarding this, Henry Morris— a professor with a Ph.D. in Civil Engineering and former head of the department at Virginia Polytechnic Institute— explains,

> Science thus involves *facts* which are observed and *laws* which have been demonstrated. The scientific method involves experimental reproducibility, with like causes producing like effects. It is *knowledge*, not inference or speculation or extrapolation.[3]

"True science," he says, "is necessarily limited to the measurement and study of *present* phenomena and processes." If this is true, then the claim of Uniformitarians and evolutionists, who say that *the present is the key to the past*, cannot be true.

Some Distinctions Within Science

Data which have been actually observed in the present, or which have been recorded by human observers in the historic past, are properly called scientific data.

"Laws, on the other hand, which have been deduced from these data, which satisfactorily correlate the pertinent data and which have predictive value for the correlation of similar data obtained from like experiments in the future, are properly regarded as scientific laws."[4] The laws of gravitation, motion and light, discovered by Isaac Newton, fit this definition of scientific laws.

Another significant distinction is between *operational science* and *historical science,* that was brought out in the nationally televised debate on origins between atheist Bill Nye and creationist Ken Ham on February 4, 2014. The essence of this distinction is as follows:

> Operational science, e.g., biology, chemistry, or physics, investigates present events that are observable, repeatable, and testable. On the other hand, historical sciences, e.g., forensics or paleontology, investigate past events that are not observable, repeatable and testable. [Bill Nye didn't seem to understand this distinction.] It does so by examining eyewitness testimony, historical records (like phone calls and emails), and remaining physical evidence of the past event. As a last resort, historical science uses presently observable processes to

hypothetically reconstruct the past and then only when current processes are sufficient to account for the evidence we have.

For example, if there is eyewitness testimony that accounts for the known facts, no judge will reject it in preference for a forensic reconstruction of how a crime may have been committed, however plausible the reconstruction may be. Credible eye-witness always trumps educated guessing.

The resurrection of Jesus is, as any "reasonable man" will admit, one of the best attested events of ancient history. Multiple, credible, eyewitness testimonies confirm it. The resurrected Son of God himself told us we should believe all that is written in the Old Testament (Luke 24:25, 44; cf. John 10:35). On the basis of Jesus' word, therefore, we have in Scripture a credible eye-witness testimony concerning the origin of the universe, and human life on earth, and the world-wide flood of Noah's day.

It is unreasonable to reject this divine testimony in favor of human guesses. Yet, that is precisely what Bill Nye and evolutionists insist on doing. Further, their evolutionary guesses aren't even scientific, for they are not based on presently observable processes, no one is observing anything even remotely close to the origin of a universe or the evolution of plant, animal, and human life.[5]

A third distinction that needs to be kept in mind is the difference between science and religion, or more precisely, the Christian religion contained in the Old and New Testaments.

The prestigious US National Academy of Science has gone on record with the following statement:

Science is a way of knowing about the natural world. It is limited to explaining the natural world through natural causes. Science can say nothing about the supernatural world [like the special creation of the world and man, the Virgin Birth of Jesus, the Resurrection of Christ, and other miracles] through natural causes. Whether God exists or not is a question about which science is neutral.[6]

But that doesn't prevent any number of scientists and some evangelical Theistic Evolutionists from weighing in with their *opinions* on supernatural events recorded in the Bible, using their vaulted positions to advance a particular theory that is in contradiction to *revealed* truth.

The Blessings of Science

True science is a gift from God. The birth of science was in the plan of God for humanity. Ironically, it started with a study of the heavens where the God of the universe dwells (Deut. 26:15; 1 Kings 8:30). As Christian scientists, Copernicus and Galileo seemed to be drawn first to astronomy because of their familiarity with the Scriptures.

The heavens declare the glory of God;
    the skies proclaim the work of his hands.
Day after day they pour forth speech;
    night after night they display knowledge. (Psalm 19:1-2)

The early scientists, like Isaac Newton (1642-1729), the father of modern physics, laid the foundation for subsequent scientists and especially for the great scientific advances of the twentieth century.

To show the great blessings of science we have only to remember that in 1900, ninety percent of the American population lived in rural and agricultural areas. By 2000, ninety percent of our population lived in urban, highly industrial and technological areas of the country. What a difference one hundred years makes!

In one century, we have gone from—

The horse and buggy .... to flashy sports cars
Kerosene lamps .... to incandescent lights
Telegraph .... to cell/smart phones
Iceboxes .... to refrigerators
Trains .... to supersonic airplanes
Hand-held fans .... to air conditioning
Washboards .... to washers and dryers
Country stores .... to supermarkets
Typewriters .... to computers
Clapboard rural churches ...... to huge city mega churches
The steam engine .... to the search engine
Exploring the Arctic Pole ...... to landing a man on the moon

These tremendous improvements in the quality of life have come about largely because of God-given human insight, research, experimentation (trial and error), the American spirit of innovation, invention, science and technology. We simply stand amazed at how God has blessed America and countries of the West in the last century.

These marked improvements and scientific advances have taken place for the most part in the Christian West. This has led some historians and sociologists to refer to this situation in our world as the "haves" (nations in the Northern Hemisphere) and the "have nots" (nations in the Southern Hemisphere.) However, it should not be overlooked, as it so often is, how much America and the West, especially since World War II, have come to the aid of the Two-thirds World with economic aid and development, medical vaccines and drugs, food and finances in times of national disaster, and efforts to spread democratic ideals and institutions around the world. That is the positive side of American goodness, compassion, and sharing the results of scientific progress.

Our material gains, however, have far outstripped our spiritual gains, especially since the Second World War. In the twentieth century, science has been a bane as well as a blessing in our lives. In modern life, the foundations of truth are science and rationalism and not the infallible Word of God. The center of modernity is man and his ability to solve all problems. Science has come to rule our lives. Even though science has improved our lives beyond anything we could have dreamed at the beginning of the century and holds sway over our culture, it is not infallible. It does not stand in the place of God; it has its limitations.

The Limitations of Science

This may be hard for some scientists to acknowledge but it is true.

First, by definition, science is a way of knowing about the natural world. It is limited to explaining the natural world through natural causes. These are the first restrictions or boundaries inherent in and placed upon modern science.

Second, true science is limited to the measurement and study of present phenomena and processes that come to the attention of the scientist.

•The science of *physics* deals with the present processes of the physical word.

•The science of *chemistry* deals with the present chemical properties and behavior of matter.

•The science of *geology* deals with the present geological processes and earth features such as sinkholes, active volcanoes, the earth's crust, plates and fault lines wherever they exist in the world. This eliminates theories of Uniformitarianism and old earth that extrapolate the age of the earth billions of years back into the past. *The present is not the key to the past.*

•The science of *biology* deals with the processes of life in plants, animals and man at the present time. This precludes any consideration of the development of organic life from non-organic life, or of the transformation of one species into another by natural selection and chance variation over long periods of time.

Henry Morris observes appropriately,

So long as the question of origins or ends is not considered, there will be no conflict between the Bible and science. The Bible has numerous references to present phenomena of science, and all will be found in strict accord with the actual observed data. It is only when questions of origins or destinies (or fundamental meanings) are considered that conflicts arise.[7]

Third, in 1954 Bernard Ramm, writing in *The Christian View of Science and Scripture*, was asked, What are the necessary limitations of evolutionary theory? He answered, "There are limits beyond which the theory of evolution may not be pushed... "

•Evolution can never become the self-creation of Nature.

•Evolution can never be the rationale of the universe.

•Evolution must reckon with energy and design in Nature.

•Evolution must face the transcendental nature of man.[8]

Contrary to the atheist and evolutionist, who do not believe in God and who believe that nature is all there is, man was created for two worlds, the physical and the spiritual; for this world and the next world. He transcends his own physical body, proving that he also has a mental and spiritual nature which must come from above and not from below. This created world has two spheres of reality and being; the physical and the metaphysical, or the natural and the supernatural.

It is certainly a joy to live in this physical world that God so graciously created for man to live in for fifty, seventy, or one hundred years of his earthly sojourn. It is also marvelous to enjoy the material blessings that twentieth-century science has provided for us. But these material things cannot bring us true happiness or satisfy the inner spirit of man. For in the beginning we were created to have fellowship with God and to enjoy Him forever. St. Augustine was right when he wrote, "The soul of man is restless until it finds its rest in God."

Thus, the study of origins, destinies, and meanings of life is properly to be considered outside the domain of science. Science has no answers to the questions, (1) Who am I?; (2) Why am I here for 30, 60, 80, or 100 years?; and (3) Where am I going when I leave this life? Only Genesis and the Bible give us these answers.

Fourth, another limitation of science is that scientists themselves are finite and fallible.

The psalmist wrote, "What is man that you are mindful of him, the son of man that you care for him? You made him a little lower than the angels and crowned him with glory and honor." (8:4-5)

The Prophet Isaiah wrote this for God: "For my thoughts are not your thoughts, neither are your ways my ways," declares the Lord. "As the heavens are higher than the earth, so are my ways higher than your ways and my thoughts than your thoughts." (Isa. 55:8-9)

Despite man's exalted position, he is still finite and fallible, and lacking in knowledge concerning nature and the universe. Only God is omniscient or all-knowing.

I like the way Marius Forté and Sam Sorbo have expressed this idea:

Science, as necessary as it is, is constantly searching to explain the unknown, and when it has discovered a small slice of new materialistic knowledge that God has laid out, like bait, for it to finally uncover, it proudly proclaims to be so much closer to Knowledge (with a capital K). And this is where science errs so profoundly. Many hundreds of years ago, scientists thought they were very close to finding the knowledge that would give them the secret to changing any metal into gold, alchemy. They also believed they could gain the knowledge of everything. They figured they just needed to open one more door and...

Eureka! To their astonishment, behind that one door were another ten doors. So, they took up the task to open those ten doors only to find a hundred more doors behind every door they opened. And behind those doors were again perhaps 10,000 more. In a way, the closer they came in their quest for knowledge, the farther they found themselves from gaining the knowledge they sought. The creator of this universe is so much greater than his creation and so much greater than even imaginable. *Science and human scientists are at a complete disadvantage.*[9] (emphasis added)

In view of these limitations of science, what should be the attitude of the scientist, especially as it relates to the universe, nature, and man? Certainly not one of arrogance and dogmatism. But one of humility, recognizing that God reveals truth and knowledge to those who are humble in spirit and have reverence for Him (Job 2:3; Prov. 1:7; Rom. 11:33)

Agreement Between Science and Scripture

What we have said to this point about science and its limitations is not meant to be disparaging in any way. Science, and its many branches, is a legitimate discipline, like any other field of study. It arose providentially in the sixteenth century as Renaissance men began to "look to the heavens" and make discoveries that were not only breathtaking, but also foundational for succeeding centuries. The climax to date has come in the twentieth century with an abundance of inventions and discoveries that have enhanced our way of life in the West. And science will continue to advance as it builds on present achievements and remains within its legitimate boundaries of present processes, that is, examining the natural world through natural causes without infringing on the supernatural or spiritual world of origins, meanings and ends (destinies).

I agree with Henry Morris who said, "In the final analysis, all truth is one. God did not create one universe of physical reality and another of spiritual reality. The same God created all things, and His Word was given by His Holy Spirit to guide us into all truth."[10]

Old Testament scholar Edward J. Young, in his *An Introduction to the Old Testament*, observes correctly,

Although Genesis does not purport to be a textbook of science, nevertheless, when it touches upon scientific subjects, [like archaeology, anthropology, astronomy, medicine, taxonomy, and others] it is accurate. Science has never discovered any facts which are in conflict with the statement of Genesis 1.[11]

Let us notice now, various areas of agreement between science and Scripture.

Creation Events (confirmed by science)

1. Time, space & matter had a beginning (Gen. 1:1)

Einstein and his theory of General Relativity have been proven to the certainty of 10.[14] This theory and the equation ($E=mc^2$) prove that matter, gravity, time, energy and acceleration are all interrelated. Applied physics has conducted many tests verifying the accuracy of this concept and equations. Among many other things, several important conclusions have been proven, namely, that time itself and space and matter all had a beginning.[12]

It is interesting that this discovery caused Einstein to switch from complete atheism to belief in a kind of creator.[13]

2. Heavenly bodies created (Gen. 1:1); this includes the solar system.

The earth was initially covered with a thick layer of gas and dust *not allowing light to penetrate*. This is probably a standard condition of planets of the earth's mass and temperature. The initial conditions described in the *Bible* are accepted by science: Dark, formless and void, or empty.[14]

3. "Let there be light" (Gen. 1:3)

Atmosphere became translucent to allow some light to reach the surface of the water and land, a critical prerequisite for the introduction of life on earth (photosynthesis).[15]

4. Development of hydrologic cycle (Gen. 1:6)

The "perfect" condition of temperature, pressure and distance from the sun would allow all forms of H_2O (ice, liquid and vapor)— all necessary for life.[16]

This may be what scientists mean when they say that the universe was precisely tuned for life, or the Anthropic Principle.[17]

The earth's atmosphere contains the right balance of oxygen and carbon dioxide (CO_2) to sustain plant, animal and human life.

5. Formation of land and sea (Gen. 1:9-10)

Seismic and volcanic activity occurred in the precise proportion to allow thirty percent of the surface of the earth to become and remain land. Scientists have determined this is the ideal ratio to promote the greatest complexity of life forms.[18]

6. Creation of vegetation (Gen. 1:11)

Light, water and large amounts of CO_2 set the stage for vegetation. This was the first life form (plant).[19]

7. Atmosphere transparency (Gen. 1:14)

Plants gradually produced oxygen to a level of 21%. This (and other factors) caused a transparent atmosphere to form and permitted "Lights in the heavens," that is, the sun and the moon, to become visible at the surface of the earth, marking seasons and days and years.

Note: Thorough understanding of *Genesis 1* requires considering the original Hebrew text. The English translation can be misleading. For example: on day four, verse 16 might imply that the sun and moon were created after the formation of plants... a problem for scientists. The actual Hebrew verb and tense used in conjunction with the words in Gen. 1:1, correctly indicate the sun and moon "became visible" at the surface of the earth on day four, but were previously created.

Therefore, it seems reasonable and safe to assume that "In the beginning" when "God created the heavens and the earth," He no doubt created the entire solar system with the sun at the center and the moon as the earth's satellite.[20] (See also Henry Morris, *Studies in the Bible and Science*, 33-34.)

8. Creation of small sea animals and birds (Gen. 1:20)

Scientists agree with the Scriptures that fish and fowl followed the plants in their appearance on the earth, preceding the land animals, and that these were the first animal life forms of all classes discussed in the Bible.[21] (See also C. Wesley King, *Creation for Earnest Believers*, 153-154).

9. Creation of land animals (Gen. 1:24)

The final life-forms prior to Man were created. These included quadrupeds— livestock and wild animals, and creatures that move along the ground like rodents.[22]

10. Creation of Man (Gen. 1:26-27);

Final creature appearing on earth.[23]

The created order in Genesis is remarkable. First, God created the heavens and the earth. It seems very probable that the entire solar system was created at this time. Then God created cosmic light, followed by the sky, dry land and seas, vegetation, the appearance of the sun and moon, the lower animals, land animals, and finally, Man himself in the image of his Maker.

It is important to note the progression in the creation story. But this progression did not happen by natural forces, that is, by any inherent power within *nature* itself as evolutionists insist. *Nature* does not possess creative power in itself. Nor is there any hint in the Genesis record that mutations and natural selection played any part in the creative process. Only God has the power to create as Jeremiah acknowledges in his prophecy in 10:12:

> But God made the earth by his power;
> He founded the world by his wisdom
> And stretched out the heavens by his understanding.

Furthermore, in biblical thought (Isaiah 55:9-10) God's word *(däbär)* and God's deed *(däbär)* are one and the same. Especially is this evident with respect to God, for his word is His creative word. To do was to speak, to speak ("Let there be... "— this phrase appears eight times in Genesis 1), was to do or carry out. God's word resulted in action, namely, the immediate yet successive creation of the world and all things in it.[24]

To show that God is the One who is doing the creating and not some evolutionary mechanism, the sacred writer of Genesis (Moses) uses the Hebrew word *bara* for the creation of the world in 1:1, animals in 1:21, and man in 1:26-27.

On the sixth day, God proceeded in orderly fashion to create the higher animals on land, and to prepare for the introduction of man, the crowning act of creation.

After describing the animals that man would need to cultivate the ground (see 1:24-25), there is a break in the narrative with the introduction of a new and totally distinct order of creation. Verses 26 and 27 declare,

Then God said, "Let us [most probably a reference to the triune Godhead] make man in our image, in our likeness, and let them rule over the fish of the sea and over the birds of the air, over the livestock, over all the earth, and over all the creatures that move along the ground." *So God created man in his own image, in the image of God he created him; male and female he created them.*

What could be plainer? Man, with all of his marvelous complexity, is not the product of millions of years of evolutionary development, *nor* does he share a common ancestry with the chimp, who is among the animals created in Genesis 1:24-25. Man *is* the handiwork of an omniscient, loving, all wise and all powerful Creator.

11. Science confirms the Scripture (Gen. 1:29-30)

This is by teaching that man and the animals are entirely dependent, directly or indirectly, upon the plant kingdom for sustenance. And both are ultimately dependent on God, the One who sustains the world and provides the rain and sunshine for plants and crops to grow.

12. No additional Creation (Gen. 2:2)

No unique creation has occurred since.

God's Perfectly Created and Finished Work

And God saw everything that he had made, and behold, it was very good, and there was evening and there was morning, the sixth day. Thus the heavens and earth were finished, and all the host of them. And on the seventh day God finished his work that he had done, and he rested on the seventh day from all his work that he had done. So God blessed the seventh day and made it holy, because on it God rested from all his work that he had done in creation. (Gen. 1:3-2:3 ESV)

The English Standard Version commentary on 1:31 reads:

Having previously affirmed on six occasions that particular aspects of creation are "good" (vv. 4, 10, 12, 18, 21, 25), God now states, after the creation of man and woman, that *everything* he has made is *very good;* the additional *behold* invites the reader to imagine seeing creation from God's vantage point. While many things do not appear to be good about the present-day world, this was not so at the beginning. Genesis goes on to explain why things have changed, indicating that no blame should be attributed to God. Everything he created was very good; it answers to God's purposes and expresses his own overflowing goodness. Despite the invasion of sin (ch. 3), the material creation retains its goodness (cf. 1 Tim. 4:4). [25]

Looking Ahead

In Part III we shall be dealing with Theistic Creationism in contrast to Theistic Evolution. Theistic Creationism is the conservative, traditional and biblical view of Genesis 1 and 2.

Endnotes

1. Henry Morris, *Studies in the Bible and Science* (Philadelphia: Presbyterian and Reformed Publishing Co., 1967), 151.

2. Loc. cit.

3. Loc. cit.

4. Loc. cit.

5. A. Philip Brown II, "Origins Debate," *God's Revivalist and Bible Advocate* (May 2014), 18.

6. Marius Forté and Sam Sorbo, *The Anwser* (Telemachus Press, 2013), 197.

7. Morris, op. cit., 153.

8. Bernard Ramm, *The Christian View of Science and Scripture* (Grand Rapids: Wm. B. Eerdmans Publishing Company, 1955), 273-280.

9. Forté and Sorbo, op. cit., 203.

10. Morris, op. cit., 120.

11. Edward J. Young, *An Introduction to the Old Testament* (Grand Rapids: Wm. B. Eerdmans Publishing Co., 1950), 54.

12. Ralph Muncaster, *The Bible: Scientific Insights* (Mission Viego, CA: Strong Basis to Believe, 1966), 21.

13. Loc. cit.

14. Ibid, 13.

15. Loc. cit.

16. Loc. cit.

17. Forté and Sorbo, op, cit., 20.

18. Muncaster, op. cit., 13.

19. Loc. cit.

20. Loc. cit.

21. Loc. cit.

22. Loc. cit.

23. Loc. cit.

24. Eugene Carpenter, "Cosmology," *A Contemporary Wesleyan Theology* (Salem Ohio: Schmul Publishing Co. Inc., 1992), 158.

25. ESV Study Bible (Wheaton, IL: Crossway, 2008), 52.

PART 3
THEISTIC CREATIONISM—
THE CONSERVATIVE/BIBLICAL VIEW
From "the Beginning" – Genesis 1:1 to the Present

9

Foundations for Interpreting Genesis 1 and 2

Overview

I N *PART 1* WE TRACED the trajectory of evolution from Charles Darwin to the current expression of Theistic Evolution represented by geneticist Francis S. Collins.

In *Part 2* we examined the origin and role of science in the modern world especially in relation to divinely-revealed truth about creation in the Scriptures. We also included a chapter on the definition and limitations of science together with points of agreement between science and the facts of creation recorded in Genesis 1.

Now, in *Part 3* we wish to present the conservative, traditional and biblical case for Theistic Creationism, which has been the belief of most Jews and Christians from "the beginning" to the present.

We will develop Part 3 in the following manner:

Chapter nine: "Foundations for Interpreting Genesis 1 and 2;"

Chapter ten: "The Interpretation of Genesis 1 and 2 Using the Inductive Method;"

Chapter eleven: "Final Conclusions based on the Examination of Genesis 1 and 2;" and

Chapter twelve: "The Authority of the Word and the Need for Centurion-like Faith."

Foundations and Buildings

Everyone will agree that foundations are extremely important in life.

One World Trade Center. Take the new One World Trade Center (WTC) in lower Manhattan, for example. Over a decade in the making, One World Trade Center is a marvel of modern skyscraper construction. Here's what went into it: 48,000 tons of steel, 208,000 cubic yards of concrete, 90% of office space is exposed to natural light, 85% of construction waste was recycled, 10,000 construction workers helped to build it, there were zero fatalities during the construction, and it was built at a cost of $4 billion. This magnificent superstructure rises 1,776 feet in the air, symbolizing the U.S.'s year of independence, making it the tallest in the Western Hemisphere and the third tallest in the world.

The building's 500,000 square foot *foundation* includes 23,000 cubic yards of concrete, enough to build a sidewalk four inches thick and four feet wide from NYC to Chicago. Underground, 27 box columns support the *foundation* of this architectural giant. What a testament to man's engineering genius and capability.[1]

The Kingdom of God and Christ's Church. In marked contrast to, and far surpassing this earthly and man-made building, is the God-implanted Kingdom of God on earth, which "is like a mustard seed, which is the smallest seed you plant in the ground. Yet when planted, it grows and becomes the largest of all garden plants, with such big branches that the birds of the air can perch in its shade" (Mark 4:31-32 NIV).

This tiny Kingdom of God that began with Jesus, the twelve Apostles and the 120 believers on the Day of Pentecost, has now spread its branches over every continent and nearly every nation, bringing spiritual redemption, shelter and fellowship to millions of weary pilgrims through the centuries.

Consequently, [we] are no longer foreigners and aliens, but fellow citizens with God's people and members of God's household, built on the *foundation* of the apostles and prophets, with Christ Jesus himself as the chief cornerstone. In him the whole *building* is joined together and *rises* to become a *holy temple* in the Lord. And in him you too are being built together to become a dwelling in which God lives by his Spirit. (Eph. 2:19-22, emphasis added)

By the grace God has given me, I laid a *foundation* as an expert builder, and someone else is building on it. But each one should be careful how he builds. For no one can lay any *foundation* other than the one already laid, which is Jesus Christ. If any man builds on this *foundation* using gold, silver, costly stones, wood, hay or straw, his work will be shown for what it is, because the Day will bring it to light. It will be revealed with fire, and the fire will test the quality of each man's work. If what he has built survives, he will receive his reward. If it is burned up, he himself will be saved, but only as one escaping through the flames. (1 Cor. 3:10-15, emphasis added).

In antiquity, the Flood destroyed men, animals and the earth because it was corrupted and full of violence (Gen. 6:7, 11-13). In the last days of human history, "By the same word [of God] the present heavens and earth are reserved for fire, being kept for the day of judgment and destruction of ungodly men" (2 Pet. 3:5-7). They will bring about the destruction of the heavens [the same heavens that God created in the beginning] by fire, and the elements will melt in the heat. But in keeping with his [God's] promise we are looking forward to a new heaven and a new earth, the home of righteousness (2 Pet. 3:12b-13).

Some day soon this judgment will fall upon the earth. Everything, including the skyscrapers that man builds, whether in New York or Dubai, will be burned up. But, the Kingdom of God will go on forever and ever.

The *foundations* of the Church of Jesus Christ have been laid. The Church (the spiritual superstructure, which is believers) will go on forever. The *foundations* of Christ's church are being challenged today as never before. "Nevertheless, God's solid *foundation* stands firm, sealed with this inscription: 'The Lord knows those who are his,' and 'Everyone who confesses the name of the Lord, must turn away from wickedness'" (2 Tim. 2:19).

What are the *foundations* that support the Bible, Genesis 1 and 2, and the ongoing work of the kingdom until Christ returns? The first foundations are biblical.

Biblical Foundations

All of the world's religions have their so-called sacred writings. Hinduism has its Vedas and Upanishads; Buddhism, its Four Noble Truths, and Islam, its Koran. And all purport to have received "divine" truth. In reality, though, Christianity and the Bible are totally unique in human history because they come from the one true God. They reveal man's true condition before God, and the real possibility of him being spiritually reunited with his Creator in this life and enjoying eternal life with Him throughout eternity. The authority of the Bible as a standard for doctrinal truth and ethical practice clearly rests upon the degree to which the Scriptures are more than a human production. The character of the Bible as the supreme revelation from God to man depends upon what is known as its "inspiration." Thus we read, "All Scripture is given by inspiration of God, and is profitable for doctrine, for reproof, for correction, for instruction in righteousness: That the man [or woman] of God may be perfect, thoroughly furnished unto all good works" (2 Tim. 3:16, 17 KJV). "Is given by inspiration" means that the Scriptures are God-breathed. This means that the eternal God of the universe is the author of all Truth; that when he breathed truth into the Scriptures, He was conveying two specific themes to humanity— Creation and Redemption— both of which are recorded in God's written Word.

Genesis is the first and foundational book of both the Pentateuch and the entire biblical record. No wonder the devil is so interested in casting doubt upon this section of the Holy Scriptures and even causing men to disbelieve its authenticity. Theologian Charles Carter claims that "Genesis is, in fact, the epitome of God's entire revelation as it relates to human beings and the created universe, over which they were commissioned by God to be the sub- sovereign custodians. Here we have a statement of the biblical account of the origin of the human race together with God's purpose and plan for that race."[2]

The word *Genesis* means "first," "beginning," or "origin." It contains the beginnings of:
1. the universe (1:1)
2. the earth (1:2-2:1)
3. the Sabbath rest (2:2-3)
4. the special creation of man (2:7)
5. the special creation of woman (2:18, 20b-25)
6. human sin and its ramifications (3:6ff; 6:1-9:17)
7. God's promise to redeem sinful man (3:15)
8. civilization and societies (4:17b)
9. nations and cultures (10:1-32)
10. human languages (11:1-9)

11. a special relationship between God and Abraham (11:1-12:9)
12. God's chosen people— the Hebrews in the Old Testament and by inference the Jews and Christians in the New Testament (12:1-50:26).

It is interesting to note that the word *Genesis* is closely related to the scientific terms *genetics* and *genes*, but not in the biological sense. Carter drives home this point.

> Remove the Genesis record, and the rest of the Bible would be meaningless. From the Genesis record all else in the Bible issues. Without it the Bible would be like a building without a foundation. With the validity of the Genesis record, the orthodox Judeo-Christian faith stands or falls. The Bible is a record of humanity's creation, Fall, and divine redemption. This entire record is explicit or implicit in Genesis.[3]

Even though Moses was not alive when the events of Genesis took place, still conservative scholarship believes that Moses wrote or at least compiled the major part, if not the whole, of Genesis using historical records (see the generations in Genesis) as had been handed down from his forefathers, and direct personal knowledge as he was moved and directed by the inspiration of God's Spirit. Henry Halley asks, "How did the writer know what happened before man appeared?" Then he answers, "No doubt God revealed to him the remote past as later the distant future was made known to the prophets."[4] After all, it is only reasonable to believe that the eternal God who made man and the world with its moral imperatives and implications would also intervene supernaturally in the accurate disclosure and writing of these early historical (primeval) events, such as the origin of man and his world, his fall into sin, and the possibility of his redemption from sin, so that the whole human race would have a reliable and trustworthy account of who the living God is, and of these prehistoric events.

The biblical foundations, then, are the first set of foundations that support the Church and the evangelical gospel. They must be accepted as inspired by God himself, authentic in what they declare, and received in their straightforward manner. They must not be tampered with or distorted by specious interpretations. The ordinary Christian, theologian or scientist must believe that God is capable of saying what He means in understandable language.

Historical Foundations

What is history? History is a branch of knowledge that records and explains past events.[5] The history spoken of here is the history of events recorded since the invention of writing. But what about historical events that took place before writing was invented, that is, in primeval history? The set of historical foundations spoken of here refer to the historicity of the people and events recorded in Genesis 1-11. Sadly, there are those today who believe that the first eleven chapters of Genesis are only legends or myths, and not historical facts and real historical events.

Recently, in a pro-evolution book from InterVarsity, *The Language of Science and Faith*, Francis Collins and co-author Karl Giberson, a physics professor at Eastern Nazarene University, escalate matters even more by announcing that "unfortunately the historical concepts of Adam and Eve as the literal first couple and the ancestors of all humans simply do not fit the evidence."[6] By evidence, Collins and Giberson mean the "near identity" of the human genome with the chimp genome and the belief (really, theory) of many biologists (including some Christian ones) that humans share common ancestry with prior primate species. But which evidence are we going to believe and use— so-called scientific, or historico-revelational?

Old Testament scholar Edward J. Young contends, "The religion presented in the Old Testament, according to its own representation, is an historical religion. It is grounded upon that which God Himself did in history. Remove this historical foundation from it and there is no longer any true biblical religion… Unless these historical facts are presupposed by faith, based on God's revelation and intervention in human history, we shall waste our time if we try to study the significance and meaning of what is narrated."[7]

The historicity of the events and people in Genesis 1-11 is borne out by these real historical facts:

- The Creator God is a *real* God in contrast to the mythological gods of the Ancient Near East.[8]
- The creation of the heavens and the earth took place in *real* time as revealed to Moses by God.

- Adam and Eve were *real* people, the first human beings specially created by God, without any predecessors or pre-Adamic people.
- Adam and Eve are mentioned as historical individuals no less than thirty times in the Scriptures (eighteen times in Genesis and twelve times in other books of the Bible by six different authors after the Genesis record: Moses, Job, Hosea, Luke, Paul, and Jude).
- The Garden of Eden was a *real* place, designed by God as man's first abode. This is confirmed by the mention of two historical rivers, the Tigris and Euphrates in Genesis 2:8-14.
- The trees in the Garden were *real* trees because God said that they could eat the fruit of them.
- They were prohibited from eating of the "tree of the knowledge of good and evil" (Gen. 2:16-17).
- "The fruit of this tree has been variously understood as giving (1) sexual awareness, (2) moral discrimination, (3) moral responsibility, and (4) moral experience. Of these possibilities, the last is the most likely; by their obedience or disobedience the human couple will come to know good and evil by experience."[9] (ESV)
- God's definition of marriage— one man and one woman— is forever and inviolable, and the only *real* concept of marriage that exists (Gen. 2:20b-24; Mark 10:6-9). All other unions are unnatural (Rom 1:24-32).
- Sin is *real* and a tragic intruder into God's plan for the human race. But there is a promised deliverer (Gen. 3:15).
- Genesis 4:17-5:32 and 10:1-32 is the *real* beginning of ethnic groups and nations, of society, the state, and human civilization. As Old Testament theologian Walter Kaiser states, "there is a *real* connection with today's world through the passage of time" (Gen. 4-11).[10]
- The flood of Genesis 7-8 was a *real*, literal and worldwide flood because of the evil and corruption of mankind.
- The scattering of mankind at the Tower of Babel in Genesis 11:1-9 over the face of the whole earth was *real* and done by God himself confusing the language of the people so they could not understand each other.

It is only reasonable to believe that the eternal God, who made man and the world with its moral imperatives, would also intervene supernaturally by divine revelation to accurately disclose and have recorded these all-important prehistoric yet real events in human history, such as the origin of man and his world, his fall into sin and the possibility of his redemption from sin by a promised Savior (Gen. 3:15). This was so that the entire human race would have a reliable and trustworthy account of who the God of the universe is and of these pre-historic events and people. We praise God that we have just such a divine record. We are not left to guess, speculate or extrapolate about who we are, why we're here, and where we're going at death.

The next foundations are interpretative.

Interpretative Foundations

In his book *Methodical Bible Study*, Dr. Robert Traina discusses some incorrect ways of interpreting the sacred Scriptures.[11]

The first is *rationalistic interpretation*. The rationalist attempts to expound the scriptures in such a way as to make them understandable and acceptable to human reason. Liberals have attributed the convulsions of the boy whom Jesus healed in Mark 9:14-29 to epilepsy rather than to demon-possession which is the clear teaching of this passage. The inability to believe certain biblical facts such as miracles often results in rationalistic interpretations.

Mythological interpretations. This approach is closely related to the preceding type in that it is often an expression of rationalism. The resurrection of Jesus is held by some not to be a real historical event, but rather a myth whose purpose it was to teach the supreme spiritual truth that though Jesus was slain, His Spirit still lives. And when one learns this important spiritual lesson, he may then discard the "story" which was used to express it.

Allegorical interpretations. This method of interpreting the Scriptures was prominent in the ancient Church among the followers of Origen (AD 180-253) and other theologians of the Alexandrian (Egypt) school of allegorical interpretation. These men, who used this method, taught that the Bible is like an onion with three layers or meanings.

•The literal meaning or outer layer corresponds to the *body*.

•The moral meaning or inner layer corresponds to the *soul*.

•The spiritual meaning or core corresponds to the *spirit*.

The current meaning of *allegorical* is "that which has hidden spiritual meaning that transcends the literal sense of a sacred text;" in this case the biblical text.

This is precisely the way Francis Collins and others use allegory in their writings. "Instead of the traditional belief in the specially created man and woman of Eden (literal meaning) who were biologically different from all other creatures, Collins mused, might Genesis be presenting 'a poetic and powerful allegory' about God endowing humanity with a spiritual and moral nature?"[12] This is exactly what Traina warns biblical interpreters not to do when he says,

> A more significant and dangerous form of allegorical explanation concerns the treatment of historical narratives [which the Genesis account of man's special creation is]. Those who use the allegorical approach frequently accept such narratives as historical, but instead of expounding their meaning in view of their concrete historical setting, they use them as allegories to teach some spiritual lessons.[13]

The vexing problem here is that Collins does not believe in the concrete historical setting, namely, the specially created man and woman of Eden. Needless to say, Collins' interpretation of Genesis 1 and 2 is absolutely wrong. It not only casts doubt on an otherwise real historical event, which is theologically foundational for the correct interpretation of Scripture as a whole, but it is dangerous in that it causes the non-Christian to continue believing in the fallibility of the Christian Scriptures.

A word needs to be said here with regard to the parable. A *parable* is a type of allegory presented in the form of a narrative in which one thing, usually a fact of nature, is placed alongside another thing, in this case a spiritual truth, for the purpose of comparison. This was Jesus' favorite method of teaching spiritual truth about the Kingdom of God that He had come to proclaim (Mark 1:14-15) and to establish in the hearts of men and women (Luke 17:21). Employed in this way, parabolic teaching is very clear, straightforward and truthful.

The Grammatical-Historical Method of Interpretation

In contrast to the rationalistic, mythological and allegorical methods of interpreting the Bible, the grammatical-historical method is by far the best and soundest method of uncovering and understanding biblical spiritual truth and applying it in daily life and culture.

First, the biblical exegete must accept wholeheartedly and without reservation the divine inspiration, integrity, inerrancy and authority of the Bible; that it is the written revelation of God's redemptive purposes for the human race promised embryonically in Genesis 3:15 and revealed progressively to holy men of God (2 Pet. 1:19-21 NKJV), prophets in the OT period of history, who spoke and wrote in Hebrew and Aramaic, until the fullness of time and the First Coming of Jesus Christ into the world (Gal. 4:4-7), and apostles in the NT period of human history, who wrote in Greek, as they were moved by the Holy Spirit (2 Pet. 1:21).

Secondly, this wholehearted belief in the inspiration, inerrancy and authority of the Bible must also include acceptance of Genesis 1-11 which is often the object of liberal and evolutionist attacks because of its seemingly mythological and unscientific nature, especially Genesis 1-3.

But, as evangelical apologist Ronald Nash reminds us, the criteria of truth by which we may test worldviews or metaphysical systems, which includes Christian theism, are: (1) consistency, (2) coherence, (3) applicability, and (4) adequacy.

> By consistency, we mean that the system must be free from internal self-contradictions... By coherence, we mean that the various points or principles of the system must stick together... Applicability means that the system [in this case Christian theism and Genesis 1-3 in particular] must be relevant to experience, that is, the worldview must be capable of illuminating some experience naturally and without distortion. Applicability means that this particular presupposition can better explain the available data more thoroughly than any other assumption.[14]

But even applicability is not enough, for the system [Christian theism] must also be adequate. Frederick Ferré, in *Language, Logic and God*, argues,

A conceptual synthesis must not only be applicable to some experience [such as human sin and salvation] which it interprets; it must (much more demandingly) be adequate to all experience [for all have sinned and fall short of the glory of God— Rom. 3:23] if it is to succeed in being of unlimited generality; that is, it must show all experience to be interpreted without oversight, distortion or "explaining away" on the basis of its key concepts.[15]

As the first and foundational book of both the Pentateuch and the entire biblical record (Genesis 1:1 to Revelation 22:21), Genesis, especially chapters 1-11, fit the above-mentioned criteria of divine truth: consistency, coherence, applicability and adequacy.

With this in mind, let's carefully examine some of the features of the grammatical-historical method of interpretation.

Grammatical principles. These principles have to do with the biblical text itself: meaning of words and sentences, structure, literary form— prose, poetry, discourse, parable, drama, apocalyptic— atmosphere of the passage. When is the language literal and when figurative?

Several fundamental rules apply as well. Since the Bible is a unity— its main theme is the redemption of mankind— scripture has only one meaning [spiritual] and should be taken literally in most instances.[16] In this regard, Ken Ham and Greg Hall's comment is *apropos,*

> Unquestionably, figurative language is used throughout Scripture to make analogies or metaphors (for example: God is our rock and our fortress) but in these passages the figure of speech is obvious. One still interprets it literally, for to do so is to consider the genre. Understanding the symbolic or figurative language enables one to literally interpret what is intended. In Genesis 1, there is no indication or any reason that the Hebrew word for "day" doesn't mean "day" in its ordinary (approximately 24 hour) meaning. Scripture makes this clear as the six days of creation are qualified by a number, evening and morning. This is a matter of authority, and it is a matter of truth— of correct interpretation. Even leading Hebrew lexicons do not leave open Genesis days as long ages:[17]
> Yôm: "Day of twenty-four hours: Genesis 1:5."[18]
> Yôm: "Day as defined by evening and morning: Genesis 1:5."[19]
> The second rule is: interpret the words in the sense that they had in the time of the author.
> The third rule is: interpret the word in relation to its sentence and to its context.

The fourth rule is: interpret the passage in harmony with its context. The fifth rule is: when an inanimate object is used to describe a living person, the proposition may be considered figurative.[20]

Historical principles. These principles have to do with the context in which the books of the Bible were written. The political, economic and cultural situations at the time of their writing are important in the consideration of the historical aspect of your study of the Word of God.

When you begin your study of a passage of scripture, imagine that you are a reporter and bombard the text with questions like:
- To whom was the letter or book written?
- What was the background of the writer?
- What was the experience or occasion that gave rise to the message?
- Who are the principal people in the book?

Several fundamental rules apply here as well. The first rule is: Since the scripture originated in a historical context, it ought to be understood in the light of biblical history.

The second rule is: Even though the revelation of God in the Scriptures is progressive in both the Old and New Testaments, they are essential parts of this revelation and form an unbreakable unity.

The third rule is: The historical facts and events become symbols of spiritual truths only if the Scriptures designate them as such.[21]

If the student of the Word will follow these clear principles of the grammatical-historical method, he or she is most likely to arrive at a sound, correct and satisfying interpretation of the passage or book under study.

The final set of foundations that support "the church of the living God" (1 Tim. 3:14-16) in this world are theological.

Theological Foundations

Theology is from the Greek *theos* and *logia,* and means "the study of God and His relation to the world that He created." Christian theology, then, is a belief system that is built on religious truth about God, man and the world revealed by God to prophets and apostles and recorded by them in the Bible, the Holy Scriptures. This is only true of the Christian religion regardless of what Muslim clerics contend with regard to their religion.

The basic biblical framework for theists, as recorded in Scripture, is built around the following key facts of history:

1) That God exists eternally;

2) That God created the heavens and the earth (Gen. 1:1; Heb. 1:10);

3) A real and special creation of all things *ex nihilo* (out of nothing) in six days, following which God stopped creating and rested on the seventh day (Gen. 2:1-3);

4) That He created human beings in the form in which they presently exist with an eternal soul (Gen. 1:26-27; 2:7, 21-25; Eccl. 3:11);

5) That He has established the rules (moral law) by which we as free and morally accountable agents to Him should play the game of life while we are here on earth for a brief time, having in view an eternal existence beyond the grave (Heb. 9:27);

6) The introduction of disobedience (rebellion), disharmony, decay and death into the world through man's fall and God's curse on the whole creation (Gen. 3; Rom. 8:18-22);

7) The destruction and renovation of the antediluvian earth and its inhabitants at the time of the great Flood (Gen. 6:1-9:17);

8) The work of redemption whereby God Himself, in the person of His Son, became flesh to reconcile the world to Himself by the substitutionary death and justifying resurrection of Jesus Christ (John 1:1-3, 14; 1 Cor. 15:3-5; 2 Cor. 5:17-21; Rom. 4:25, 5:12-21); and

9) The consummation of God's purposes for the world when Christ returns, involving wrath and judgment for all who have rejected Him, signaling the end of human history and the creation of a new heaven and earth, wherein dwells righteousness as the eternal dwelling place of the redeemed (2 Pet. 3:3-16).

This basic framework of young earth and metaphysical history is emphatically rejected, in every part, by ancient and modern intellectuals and evolutionists (scholars and scientists) and compromised, in part, by Emerging Church theologians.

The cover story in the June 2011 issue of *Christianity Today* was "The Search for the Historical Adam." Some scholars believe genome science *casts doubt* on the existence of the first man and woman (Adam and Eve). Other scholars say the integrity of the Christian faith requires it, that is, belief in the biological "evidence" discovered recently by Dr. Francis Collins and his team of researchers at the National Institute of Health, of the "near identity" of the human genome with the chimp genome, and that this is "proof" that man, as we know him today, is not a special creation of God, but rather shares common ancestry with prior primate species.[22]

Sadly, some Bible-believing professors and scientists in well-known Christian colleges, universities and seminaries have been drawn into this debate and have taken the position that the Christian faith requires belief in this so-called new biological "evidence" that man is descended from prior primate species, as the logical answer to the age-old question of how God created the first human beings. It is true that Gen. 2:7 does not give us the details as to precisely *how* God created the first human male in His image, but God has given us enough trustworthy information for us to believe that He did, without searching for some sort of biological and evolutionary evidence to prove that that is the way God created the first human male. There is simply no evidence in Scripture for the hypothesis of man's descent from a common ancestor. For evangelical scholars and scientists to take such a position or to even entertain it is not only unwarranted but dangerous. Michael Cromartie, the evangelicalism expert at Washington's Ethics and Public Policy Center, sees high stakes, calling the new thinking "urgent" and "potentially paradigm-shifting" development with "high theological implications... How this gets settled is extremely important.[23]

The stakes in this controversy are indeed very, very high. In previous decades, theologically liberal Christians, atheists and evolutionists have gone to great lengths to undermine the authority and trustworthiness of the Scriptures. But now, those who should be holding the line on the complete integrity of the Scripture are toying with or even believing in a different origin of man. This theory not only goes to the heart of the Christian belief of who man really is, but it goes to the heart of the Gospel itself. This is why I call this "The Battle for Genesis 1 and 2." Satan, as the archdeceiver and enemy of God, man, and the Bible, is now "going for the jugular" of Christianity, because he sees his end approaching. The future of evangelicalism hangs in the balance. Which worldview are we going to cling to— some new-fangled biological discovery or the revealed truth in Genesis 1 and 2?

Chapter ten will be devoted to an analysis and interpretation of Genesis 1 and 2 using the inductive method of interpretation.

Endnotes

1. Josh Sanburn, "America's Tower" *Time* (New York: Time Inc., 2014), Insert.

2. Charles Carter, "Anthropology," in *A Contemporary Wesleyan Theology*, (Grand Rapids: Salem, OH: Schmul Publishing Co., Inc., 1992), 202.

3. Loc. *cit.*

4. Henry H. Halley, *Halley's Bible Handbook* (Grand Rapids: Zondervan Publishing House, 1965), 59.

5. *Webster's Ninth New Collegiate Dictionary* (Springfield, MA: Merriam Webster, Inc., Publishers. 1986) 573.

6. Richard Ostling, "The Search for the Historical Adam" in *Christianity Today*, June 2011, 24.

7. Edward J. Young, *The Study of Old Testament Theology Today* (Westwood, NJ: Fleming H. Revell Company, 1959), 15, 30.

8. See my book *Creation for Earnest Believers*, 19-21. This is a self-disclosed fact embedded in human history.

9. *ESV Study Bible* (Wheaton, IL: Crossway, 2011), 54.

10. Walter C. Kaiser, *Toward an Old testament Theology* (Grand Rapids: Zondervan Publishing House, 1978), 68.

11. Robert A. Traina, *Methodical Bible Study* (No Publisher indicated, 1968), 169-175.

12. Ostling, op. cit., 24.

13. Traina, op., cit., 172.

14. Ronald H. Nash, *The New Evangelicalism* (Grand Rapids: Zondervan Publishing House, 1963), 119.

15. Frederick Ferré, *Language, Logic and God* (New York: Harper and Row, 1961), 163.

16. Walter A. Henrichsen, *Princípios de Interpretacão da Bíblia* (São Paulo: Editora Mundo Cristão, 1980), 36.

17. Ken Ham and Greg Hall, *Already Compromised* (Green Forest, AR: Master Books, 2011), 124.

18. Ludwig Koehler and Walter Baumgartner, *Hebrew and Aramaic Lexicon of the Old Testament*, Volume 1 (Leiden; Boston, MA: Brill, 2001), 399.

19. Francis Brown, S.R. Driver, and Charles A. Briggs, *Hebrew and English Lexicon of the Old Testament*, 9th printing (Peabody, MA: Hendrickson Publishers, 1906), 298.

20. Henrichsen op. cit., 39, 42, 44, 46.

21. Ibid, 56, 58, 60.

22. Ostling, op. cit., 23, 25.

23. Ibid, 24.

10

Genesis 1 and 2: What Does It Say and What DoesIt Mean?

Some Perspective

T HE INTRODUCTION TO the Book of Genesis in the Life Application Study Bible begins this way:

We sometimes wonder how our world came to be. But here [in Genesis 1:1-2:4] we find the answer. God created the earth and everything in it, and made humans like himself. Although we may not understand the complexity of just how he did it, it is clear that God did create all life. This shows not only God's authority over humanity, but his deep love for all people... The Bible does not discuss the subject of evolution. Rather, its worldview assumes God created the world. The biblical view of creation is not in conflict with science; rather, it is in conflict with a worldview that starts without a creator. Students of the Bible and of science should avoid polarizations and black/white thinking. Students of the Bible must not make the Bible say what it doesn't say, and students of science must not make science say what it doesn't say.[1]

While I am in agreement with the above statement, what does the writer mean by "polarizations"? Does he mean that there is some kind of "happy median" between what the Bible says in Genesis 1 and 2 and what science may have to contribute to the discussion of this passage? If that is what he means, there is an objection. Science and the Bible are not of equal value when it comes to interpreting what Genesis 1 and 2 says, particularly what it means. Science and the Scriptures are in two separate domains or provinces, as pointed out in chapter eight, where science was defined, including some of its limitations.

Francis Collins agrees with me on this point, that these are two separate realms. In the introduction to his book, *The Language of*

God, he says, "Science's domain is to explore nature. God's domain is in the spiritual world, a realm not possible to explore with the tools and language of science."[2] As we know, the Bible speaks clearly about the creation, sin and redemption of man by God's grace. This is the spiritual world of which Collins speaks. Speaking of the scientific world, Henry Morris observes, "The science of biology deals with the processes of life in plants, animals and man. So long as the question of *origins* or *ends* is not considered, there will be no conflict between the Bible and science... it is only when questions of origins or destinies (or fundamental meanings) are considered that conflicts appear."[3] And this is exactly where the present crisis in the evangelical world and the current debate between Biblical Creationists and Theistic Evolutionists arose recently.

The present crisis in evangelical thinking with regard to Genesis 1 and 2 and the origin of the universe, and in particular of Adam and Eve as the first specially and divinely-created human beings, arose because of the complete mapping of the human and chimp genomes, and their near match, and the possibility that man had a different origin altogether, namely "that anatomically modern humans emerged from primate ancestors perhaps 100,000 years ago— long before the apparent Genesis time-frame— and originated with a population that numbered something like 10,000, not two individuals."[4] Francis Collins and kindred Christian biologists and scholars in a number of evangelical schools of higher education believe that the clear divinely-revealed truth of Genesis 1 and 2 regarding the origin of man must now be bent and made to conform to the "new scientific data" that Theistic Evolutionists have come up with. This is a clear intrusion of science into the spiritual domain and it cannot go unchallenged. The stakes are too high.

Therefore, in this chapter, I propose to analyze Genesis 1 and 2 using the inductive method. The inductive method is similar to the scientific method in that both seek to gather facts about a certain belief (or theory), test or analyze it, and then draw conclusions about the trustworthiness of the belief. The inductive method of Bible study is comprised of three steps: (1) Observation (what does the text say?); (2) Interpretation (what does the text mean?); and (3) Conclusions/

application. In my analysis I will be aided by a number of conservative orthodox evangelical scholars and the following reference works:

ESV—	English Standard Version, Study bible
NAS—	New American Standard. The Ryrie Study Bible
NLT—	New Living Translation, Life Application Study Bible
HGKW—	Hebrew-Greek Key Word Study Bible
RGSB—	Reflecting God Study Bible
TWB—	The Wesley Bible – A Personal Study Bible for Holy Living
NSRE—	The New Scofield Reference Edition
BSI—	The Bible: Scientific Insights.

Hermeneutical Principles Essential for a Sound Interpretation of Scripture

As we proceed to interpret Genesis 1 and 2, here are the established rules that will guide the study:

1. The inductive method does not insist that the Bible is a scientific book, nor concede that it is unscientific; it regards it as non-scientific. To call it a "scientific" book is to misjudge its genius and purpose. Although Genesis does not purport to be a textbook of science, nevertheless, when it touches upon scientific subjects, it is accurate because it was inspired by the God who has full knowledge of all scientific facts. Science has never discovered any facts which are in conflict with the statements of Genesis 1... not for an instant can its accurate statements be regarded as out of harmony with true science.[5]

2. It is best to allow the biblical writers to speak for themselves as they were inspired by the Holy Spirit to receive and record divinely-revealed truth inaccessible otherwise to human reason.

3. It is best to compare Scripture with Scripture. The Bible is its own best interpreter. For example, the New Testament is the fulfillment of the Old Testament (Ex. 20:8-11 explains Gen. 2:2-3). But more importantly and significantly, Paul's teaching in Rom. 5:12-21 about the second Adam links the historical Adam of Gen. 1-3ff with redemption through Christ.

4. Students of the Bible must not make the Bible say what it doesn't

say. We shall be very careful to avoid this. On the other hand, students of science must not make science say what it doesn't say, nor should they try to make Genesis 1 and 2 say what they think it should say and mean from their scientific perspective and bias. For this reason we will not entertain scientific speculation or theories as to the creation of the universe and man. All preconceived ideas as to how God created must be "checked at the door" before entering.

In the following inductive study of Genesis 1 and 2 we will give the scriptural reference (what it says) first. That will be followed by an explanation (what it means). An abbreviation, such as ESV, indicates the source from which the meaning is drawn.

What it Says and What it Means

Genesis 1

What it says— (1:1) "In the beginning God created the heavens and the earth."

What it means— "In the beginning" does not mean of eternity, but of time. This is the first event, the origin of the heavens and the earth (sometime before the first day), including the creation of *matter, space,* and *time.*

Heb. 11:3 and Rev. 4:11 confirm that creation was from nothing *(ex nihilo,* ESV)

The Hebrew verb and tense used in Gen. 1:1 correctly indicate the sun and moon became visible at the surface of the earth on day four. It is reasonable to believe then that the sun and moon were created, and probably the entire solar system, when God created the heavens and the earth (ESV/BSI).

What it says— (1:2) "Now the earth was formless and empty, darkness was over the surface of the deep, and the Spirit of God was hovering over the waters."

What it means— The Hebrew construction of verse two is disjunctive, however, describing the result of the Creation described in verse one. The term *bohû* translated "empty," is found only in three other places (Isa. 34:11; 45:18; Jer. 4:23). It does not describe chaos, but rather desolation or emptiness. The remainder of the chapter describes God's filling of the empty earth (HGKW).

There is no reason to postulate that a long time elapsed between Gen. 1:1 and 1:2, during which time the earth became desolate and empty (ESV). This is an argument against the so-called Gap Theory. These conditions existed between the absolute creation of v. 1 and the first creative word in v. 3.

What it says— (1:3) "And God said, 'Let there be light,' and there was light."

What it means— In Genesis 1 the absolute power of God is conveyed by the fact that he merely speaks and things are created. Each new section of the chapter is introduced by God speaking... This is the first of the ten words of creation in chapter 1 (ESV).

It is a scientific fact that light is the basic prerequisite for all physical life. It assists with photosynthesis (TWB).

What it says— (1:4) "God saw that the light was good, and he separated the light from the darkness."

What it means— The first of three separations. Here, light from darkness; then sky from water (v. 7); and finally, the land from the seas (v. 9). Only when this special separation was complete did God pronounce everything good (v. 10, NAS).

What it says— (1:5) God called the light "day" and the darkness he called "night." And there was evening and there was morning— the first day.

What it means— The focus in v. 5 is on how God ordered time on a weekly cycle (ESV). Evening and morning cannot be construed to mean *age,* but only a day; everywhere in the Pentateuch the word *day,* when used (as here) with a numerical adjective, means a solar day (now calibrated as twenty-four hours; NAS).

The whole day connoted the end of daytime (evening) and the end of nighttime (morning); see vv. 8, 13, 19, 23, 31. Evidently a twenty-four-hour day is in view (RGSB).

The Literal Day Theory holds that the universe was created in six literal days with no indefinite periods of time in between. Attempts to link the biblical account of creation with evolution are not supportable because the order of creation contradicts what would be scientifically necessary for an evolutionary process involving eons of time (HGKW). This is an argument against both evolution and the Intermittent Day Theory.

What it says— (1:6-8) "And God said, 'Let there be an expanse between the waters to separate water from water.' So God made the expanse and separated the water under the expanse from the water above it. And it was so. God called the expanse 'sky.' And there was evening, and there was morning— the second day."

What it means— Expanse here means sky, dome, canopy and atmosphere for the earth (RGSB, NAS, ESV).

Water is simply something God created, and it serves as material in the hands of the sole sovereign Creator (ESV).

Apparently God suspended a vast body of water in vapor form over the earth, making a canopy that caused conditions on the earth to resemble those inside a greenhouse. This may account for the longevity of human life (Gen. 5) and for the tremendous amount of water involved in the worldwide flood (Gen. 6-9, NAS).

What it says— (1:9-10) "And God said, 'Let the water under the sky be gathered in one place, and let dry ground appear.' And it was so."

What it means— God called the dry ground "land, and the gathered waters he called seas." And God saw that it was good.

What it says— (1:11-12) "Then God said, 'Let the land produce vegetation: seed-bearing plants and trees on the land that bear fruit with seed in it, according to their various kinds.' And it was so. The land produced vegetation; plants bearing seed according to their kinds and trees bearing fruit with seed in it according to their kinds. And God saw that it was good. And there was evening, and there was morning— the third day."

What it means— Dry land and seas in vv. 9-10 are the last two objects to be specifically named by God. God then instructs the earth to bring forth vegetation (vv. 11-12) that could not grow without light and water (ESV, TWB).

This is the beginning of organic or diffused life. God spoke these plants and trees into existence and it took place immediately. The fact that these plants and trees were seed-bearing and had seed in them means that they were mature from their creation and began growing and reproducing immediately.

The RGSB comments that their creation and reproduction are orderly.

The Ryrie S. B. admits that it is impossible to know whether "kind" is to be equated with families, genera, or some other category of biological classification. However, it does point out that there are fixed boundaries beyond which reproductive variations cannot go (NAS).

What it says— (1:14-19) "And God said, 'Let there be lights in the expanse of the sky to separate the day from the night, and let them serve as signs to mark seasons and days and years, and let them be lights in the expanse of the sky to give light on the earth." And it was so. God made two great lights— the greater light to govern the day and the lesser light to govern the night. He also made the stars. God set them in the expanse of the sky to give light on the earth, to govern the day and the night, and to separate light from darkness. And God saw that it was good. And there was evening and there was morning— the fourth day."

What it means— The ESV speaks with great clarity when it says, "The term made (Heb. asah, v. 16), need not mean that God 'fashioned' or 'worked on' them; it does not of itself mean that they did not exist in any form before this [see 1:1 above]. Rather, the focus there is on the way in which God has ordained the sun and moon to order and define the passing of time according to his purposes" (ESV).

"Serve as signs." The orderliness of time (the annual, seasonal, and daily cycles) is a gift from the Creator (RGSB).

What it says— (1:20-23) "And God said, 'Let the water teem with living creatures, and let birds fly above the earth across the expanse of the sky.' So God created the great creatures of the sea and every living and moving thing with which the water teems, according to their kinds, and every winged bird; according to its kind. And God saw that it was good. God blessed them and said, 'Be fruitful and increase in number and fill the water in the seas, and let the birds increase on the earth.' And there was evening and there was morning— the fifth day."

What it means— Having previously described the creation of the *waters* and the *expanse of the heavens,* this section focuses on how they are filled with appropriate creatures of different kinds. As reproducing organisms they are blessed by God so that they may be fruitful and fill their respective regions (ESV).

The birds and beasts were formed from materials taken from the earth (v. 21, see also 2:19). God created the earth to be inhabited. He could have filled it at once by His spoken word, but He chose to empower all living things to procreate "according to their kinds" (TWB).

The term for great sea creatures (Heb. *tannin)* in various contexts denotes large serpents, dragons, or crocodiles, as well as whales or sharks (the probable sense here). Some have suggested that this could also refer to other extinct creatures such as dinosaurs (ESV).

What it says— (1:24-25) "And God said, 'Let the land produce living creatures according to their kinds: livestock, creatures that move along the ground, and wild animals, each according to its kind.' And it was so. God made the wild animals according to their kinds, the livestock according to their kinds and all the creatures that move along the ground according to their kinds. And God saw that it was good."

What it means— Livestock, or cattle, that are large domesticated quadrupeds. Creeping things, that is, creatures that move on the earth or close to it, having no legs or, at best, only short ones (that is, worms, insects, and reptiles, NAS).

These terms group the land animals into three broad categories, probably reflecting the way nomadic shepherds would experience them: the domesticated stock animals (e.g., sheep, goats, cattle, and perhaps camels and horses); the small crawlers (e.g., rats and mice, lizards, spiders); and the larger game and predatory animals (e.g., gazelles, lions). The list is not intended to be exhaustive (ESV).

What it says— 1:26-31 "Then God said, 'Let us make man in our image, in our likeness, and let them rule over the fish of the sea and the birds of the air, over all the creatures that move along the ground.' So God created man in his own image, in the image of God he created him; male and female he created them. God blessed them and said to them, 'Be fruitful and increase in number; fill the earth and subdue it. Rule over the fish of the sea and the birds of the air and over every living creature that moves on the ground.' Then God said, 'I give you every seed-bearing plant on the face of the whole earth and every tree that has fruit with seed in it. They will be yours for food. And to all the beasts of the earth and all the birds of the air and all the creatures that move on the ground— everything that has the breath of life in it— I give every green plant for food.' And it was so. God saw all that he had made, and it was very good. And there was evening and there was morning— the sixth day."

What it means— This (1:24-31) is by far the longest section given over to a particular day, indicating that day six is the peak of interest for this passage. The final region to be filled is the dry land or earth (as it has been designated in v. 10). Here a significant distinction is drawn between all the living creatures that are created to live on dry ground, and human beings. Whereas vv. 24-25 deal with the "living creatures" that the earth is to bring forth, vv. 26-30 concentrate on the special status assigned to humans (ESV).

"Let us" is perhaps a reference to the Trinity. The Hebrew word here of man is *'adam,* to be understood as "humankind" (and so in vs. 27; 2:5, 7, 8, 15, 16, 18, 25). It is the same word used for the proper name "Adam" in 2:19 and subsequent verses. *'Adam* is related to the Hebrew word for ground, *'adamah,* from which humankind was "formed" (2:7).

Man was made in the "image" and "likeness" of God. This image is found chiefly in the fact that man is a personal, rational and moral being. While God is infinite and man finite, nevertheless man possesses the elements of personality similar to those of the divine Person: thinking, feeling, willing. Man was created, not evolved. This is expressly declared, and the declaration is confirmed by Christ (Matt. 19:4); it is also confirmed by the unbridgeable chasm between man and beast; the highest beast has no God-consciousness (religious nature, NSRE).

In the beginning, before the Fall, Adam and Eve have the moral image of God as well, which is righteousness and holiness (Eph. 4:24, RGSB).

Dominion over all of Creation (v. 28; see Psa. 8:4-8) is the functional result of bearing God's image. Humans are to be God's visible representation, ruling Creation as God would rule it. In addition, the image of God points to their distinctive dignity— endowed with reason, moral self-consciousness, freedom of choice, imagination, immortality, and at least limited creativity (TWB).

This "image" and this dignity apply to both "male and female" human beings. At this stage, humanity is set apart from all other creatures (vv. 26-27) and crowned with glory and honor as ruler of the earth (see Psa. 8:58, ESV).

"Be fruitful and multiply, increase in number; fill the earth..." This means that God's creation plan is that the whole earth should be populated by those who know him and who serve wisely as his representatives (ESV).

"Subdue it and have dominion." Here the term subdue (Heb. *kabash)* means that the man and woman are to make the earth's resources beneficial for themselves, which implies that they would investigate and develop the earth's resources to make them useful for human beings generally. This command provides a foundation for wise scientific and technological development; the evil uses to which people have put their dominion comes as a result of Genesis 3 (ESV).

This is the divine Magna Carta for all true scientific and material progress. Man began with a mind that was perfect in its finite capacity for learning, but he did not begin knowing all the secrets of the universe (NSRE).

No divine permission is given here for the shedding of blood to sustain human life. Only after the Flood did God give the animals to the man and woman for food (9:2, 3, TWB).

People and animals seem to be portrayed as originally vegetarians (vv. 29-30, RGSB).

What it says— (1:31) "God saw all that he had made, and it was very good. And there was evening, and there was morning— the sixth day."

What it means— Having previously affirmed on six occasions that particular aspects of creation are "good" (vv. 4, 10, 13, 18, 21, 25), God now states, after the creation of man and woman, that everything he has made is very good... Everything he created was very good: it answers to God's purposes and expresses his own overflowing goodness. Despite the invasion of sin (ch. 3), the material creation retains its goodness (ESV).

Genesis 2

What it says— (2:1-3) "Thus the heavens and the earth were completed in all their vast array. By the seventh day God had finished the work he had been doing; so on the seventh day he rested from all the work. And God blessed the seventh day and made it holy, because on it he rested from all the work of creating that he had done."

What it means— These verses bring to a conclusion the opening section of Genesis by emphasizing that God has completed the process of ordering Creation (ESV).

On the seventh day the universe had been finished by God. But a concluding act follows: God "rested," Hebrew *shabath,* a term meaning "to cease." God did not rest from weariness, but ceased from His creative work and blessed and sanctified, or set apart, the seventh day as a day different from the previous six days. The rationale for the fourth commandment (see Ex. 20:8-11), to "do no work" on "the seventh day... Sabbath" is based on this seventh day of creation (TWB).

What it says— (2:4-4:26) This passage is too lengthy to include here. However, we will focus on the most important verses in this section in our ongoing interpretation.

What it means— The beginning of human history is described including: a more detailed restatement of the first male (Adam) and female (Eve); the serpent's enticement of Eve to disobey God, and God's resulting curse upon the serpent (Satan) and the ground, the punishment of Adam and Eve; the birth of the first human offspring; Cain's murder of his brother Abel and God's resulting curse upon Cain, and the birth of Seth and Enoch, men who began to call on the name of the Lord (4:26, TWB).

Earth's First People centered initially on the Garden of Eden. The episodes that make up this part of Genesis recount how God's ordered Creation is thrown into chaos by the human couple's disobedience. The subsequent story of Cain and Abel and then Lamech (ch. 4) show the world spiraling downward into violence, which precipitate the Flood (6:11). These events are very significant for understanding not only the whole of Genesis but all of the Bible (ESV).

What it says— (2:4-7) This is the account of the heavens and the earth when they were created. When "the Lord God made the earth and the heavens... the Lord God formed the man from the dust of the ground and breathed into his nostrils the breath of life, and the man became a living being."

What it means— "These are the generations [account] of the heavens and the earth..." This is the first of eleven such headings that give structure to the Book of Genesis (see 5:1, which varies slightly; 6:9; 10:1; 11:10; 11:27; 25:12; 25:19; 36:1; 36:9; 37:2. Each heading concentrates on what comes forth from the object or person named (ESV).

Throughout 1:1-2:3 the generic word "God" was used to denote the deity [thirty-two times] as transcendent Creator. The reader is now introduced to God's personal name, "Yahweh," or Lord, and used together, "Lord God," they underlie the personal and relational nature of God as the covenant-making and covenant-keeping God with mankind (ESV).

What it says— See above.

What it means— These verses concentrate on God's creation of a human male, amplifying 1:26-31 in particular. The main action here is God's "forming" of the man 2:7 (ESV).

The verb "formed" (Heb. *yatsar)* conveys the picture of a potter's meticulous work in shaping a vessel from clay into a particular shape. The close relationship between the man and the ground is reflected in the Hebrew words used to denote them, *'adam* and *'adamah',* respectively (ESV).

"...And God breathed into his nostrils the breath of life" (v.7). Here God breathes life— physical, mental, and spiritual— into the one created to bear his image. Life springs directly from God. "...and man became a living being." The same term is used in 1:20, 24 to denote sea and land creatures. While human beings have much in common with other living beings, God gives humans alone a royal and priestly status and makes them alone "in his own image" (1:27, ESV).

What it says— (2:8-9) "Now the Lord God had planted a garden in the east, in Eden; and there he put the man he had formed... In the middle of the garden were the tree of life and the tree of the knowledge of good and evil."

What it means— Apparently somewhere in Mesopotamia (modern Iraq), since two of the four rivers are the well-known Tigris and Euphrates (see v. 14, NAS).

Tree of life: A sign to humans of their continuing life (see 3:22) upon condition of their continued obedience (see Rev. 22:14). Tree of... knowledge: a sign to humans of the certainty of death if they disobeyed (see v. 17) in spite of the serpent's denial (see 3:4, TWB).

What it says— (2:15-17) The Lord God took the man and put him in the Garden of Eden to work it and take care of it. And the Lord God commanded the man, "You are free to eat from any tree... but you must not eat from the tree of the knowledge of good and evil, for when you eat of it you will surely die."

What it means— Adam was a farmer, zoologist and caretaker of Eden.

The fact that the command was given to Adam implies that God gave "the man" a leadership role— a role that is also related to the leadership responsibility of Adam for Eve his wife (see v. 18). For the NT understanding of the relationship between husband and wife, see Eph. 5:22-33 (ESV).

The fruit of the tree of the knowledge of good and evil is more than likely moral experience. By their obedience or disobedience the human couple will come to know good and evil by experience. Experience gained by "fearing the Lord" (Prov. 1:7) is wisdom, while that gained by disobeying God is slavery (ESV).

Through Adam's disobedience both moral and physical death would infect humanity throughout all generations. But God's plan of salvation [hinted at in Gen. 3:15] includes the obedience of the Man, Jesus Christ, through whom we may gain eternal life (see Rom 5:12-21 and 1 Cor. 15:21-22, TWB).

What it says— (2:18-25) "The Lord God said, 'It is not good for the man to be alone. I will make a helper suitable for him.' ...So the man gave names to all the livestock, the birds of the air and all the beasts of the field. But for Adam no suitable helper was found... then the Lord God made a woman from the rib he had taken out of the man, and he brought her to the man."

What it means— These verses describe how God provides a suitable companion for the man (see vv. 18, 20b). In order to find the man a helper fit for him, God brings to him all the livestock, birds and beasts of the field. None of these, however, proves to be "fit for" the man (ESV).

Loving companionship was part of God's intentions for the man as a human being. However, a partner who would provide such companionship was not to be found among the animals, but only in the creation of woman (TWB).

When no suitable companion is found, among all the living beings, God fashioned a woman from the man's own flesh (ESV).

The sentence "this at last is bone of my bone and flesh of my flesh" and the story of Eve's creation both make the point that marriage creates the closest of all human relationships. It is also important to observe that God creates only one Eve for Adam, not several Eves or another Adam. This points to heterosexual monogamy as the divine pattern for marriage that God established at creation (ESV).

Charles Ryrie reinforces this when he says, "This verse (v. 24) emphasizes the complete identification of the two personalities in marriage. The passage tells us that God instituted marriage and that it is to be monogamous, heterosexual, and the complete union of two persons. Jesus added that it is to be permanent (see Mark 10:7-9, NAS).

Because God took woman from man (see v. 21) they were originally one flesh. To become again one flesh is figurative for the physical act of sexual union, intended by God to take place only within marriage (TWB).

What it says— (2:25) "The man and his wife were both naked, and they felt no shame."

What it means— This final description in vv. 18-25 offers a picture of innocent delight and anticipates further developments in the story. The subject of the couple's nakedness is picked up in 3:7-11 after their disobedience to God's command and their eating of the forbidden fruit (ESV).

Genesis 3

What it says— (3:7-11) "Then the eyes of both of them were opened, and they realized they were naked; so they sewed fig leaves together and made coverings for themselves... And he [God] said, "Who told you that you were naked? Have you eaten from the tree that I commanded you not to eat from?"

What it means— "Knew...they were naked": A figure for shame. "Made... coverings:" Shame always seeks to conceal. Prior to their disobedience the man and the woman had experienced no guilt in their relationship with each other or with God (see 2:25). "Hid themselves:" When we have broken fellowship with God through sin, He is always seeking to draw us back to Himself (TWB).

The disastrous consequences of Adam's sin cannot be overemphasized, resulting in the fall of mankind, the beginning of every kind of sin, suffering and pain, as well as physical and spiritual death for the human race (ESV).

Conscious of the Lord God's presence, they hid. When confronted by God regarding the tree of the knowledge of good and evil, the man blames the woman, who in turn, blames the serpent (ESV).

What it says— (3:15) "And I [God] will put enmity between you [the serpent, figuratively meaning Satan] and the woman, and between your offspring and hers; he will crush your head and you will strike his heel."

What it means— This is one of the most important verses in the Bible because it contains the first veiled promise of the Redeemer, Jesus, the seed of Mary. In His death on the cross Jesus' heel would be bruised by Satan, but in His resurrection Jesus would destroy the power of Satan forever (ESV).

This verse has traditionally been understood as pointing forward to the defeat of the serpent by a future descendant of the woman. This defeat is implied by the serpent being bruised in the head, which is more serious than the offspring of Eve being bruised in the heel. For this reason, v. 15 has been labeled the *Protoevangelium,* the first announcement of the gospel— "good news" of salvation through faith in Jesus Christ (ESV).

This is the Word of the Lord, and this is the one, true interpretation of Genesis 1-3.

Closing Statement

In a thoughtful and reverent way, Old Testament scholar Eugene Carpenter offers this closure to our induction study of Genesis 1 and 2:

Genesis 1-2 must not be read merely as a straightforward declaration of God's objective creation, but also as a religious— a theological— statement that conveys to faith the origin of all things (Heb. 11:3); not merely the existence of the atom, but the basis for the existence of morals, values, religion— for the world as we know it. Any account of the origin of all things should provide the basis for a holistic *interpretation* of the cosmos. Genesis 1-2 does this well.[6]

Endnotes

1. *Life Application Study Bible* (Wheaton, IL: Tyndale House Publishers, Inc., 1996), 5.

2. Francis Collins, *The Language of God* (New York: Free Press, 2006), 6.

3. Henry Morris, *Studies in the Bible and Science* (Philadelphia: Presbyterian and Reformed Publishing Co., 1967) 153.

4. Richard Ostling, "The Search for the Historical Adam," *Christianity Today*, June 2011, 24.

5. Edward J. Young, *An Introduction to the Old Testament* (Grand Rapids: Wm. B. Eerdmans Publishing Co., 1950) 54.

6. Eugene Carpenter, "Cosmology," in *A Contemporary Wesleyan* Theology (Salem, OH: Schmul Publishing Co., In., 1992), 155.

11

Conclusions Based on an Analysis of Genesis 1 and 2

Introduction

A S NOTED IN CHAPTER TEN, the inductive method of Bible study is comprised of three steps: 1) *Observation* (what does the text say?); (2) *Interpretation* (what does the text mean?); and (3) *Conclusion(s)/Application* to one's personal life and surrounding culture. In one sense, how we apply what we learn from the Scriptures is the most important of these three steps. The Apostle James, half-brother our Lord, certainly thought so (see James 1:22-25; 2:14-26).

On the other hand, the interpretation of Scripture is just as important as drawing conclusions and applying biblical teachings in our lives as followers of Christ. And certainly correct interpretation of the Word is *all*-important. For example, in our day, there is a renewal of *universalism*. Thom Rainer notes that a few of the most liberal denominations and churches have responded to postmodernism with a message of universalism, the belief that in the end all will be saved.[1] This belief is based on an incorrect interpretation of God's Word and a faulty understanding of God's nature. God is love, but He is also judge of all mankind.

A more common response to the postmodern spirit is *pluralism*, a belief that many religions or belief systems have claims to truth. Some claim that, while Christian tenets are true, so are the positions of Islam, Mormonism and Buddhism. All "good" religions contain truth and stand alongside of each other with equal and valid claims.[2]

But this cannot be true, because only the Son of Man/God came into this world to seek and to save what was lost (Luke 19:10), and all of us are lost. Before giving His life on the cross for our salvation, the Disciples asked, "Lord, we don't know where you are going, so how can we know the way?" Jesus answered, "I am the way and the truth and the life, no one comes to the Father except through me'" (John 14:5-6). Then, after the Day of Pentecost, Peter boldly declared, "Salvation is found in no one else, for there is no other name under heaven given to men by which we must be saved" (Acts 4:12).

The Bible is the only book that tells us about the human condition, and redemption from sin, and the Fall, about restoration to fellowship with God by grace, and regaining the moral image of God, which is righteousness and true holiness (Eph. 4:24). So, how we treat (interpret) the opening chapters of Genesis, especially as they relate to the first two human beings that God created, is extremely important.

In our analysis of Genesis 1 and 2 in the previous chapter, we brought the best of conservative, evangelical scholarship to bear upon the crucial issues in these foundational chapters of Genesis. These scholars have done us an invaluable service in bringing out what these verses really mean based on a correct interpretation of the original Hebrew text communicated to Moses by God the Holy Spirit. They have also given diligent attention to the original text using sound hermeneutical principles. The conclusions they have come up with and offer for our full consideration are as follows:

The Literal Meaning of "Day" in Genesis

Biblical critics of Genesis 1 have long held that the meaning of *yôm,* the Hebrew for "day" in this chapter, refers to an indefinite period of time rather than an ordinary day of twenty-four hours. But our evangelical scholars unanimously agree that this is simply not true (see ESV, NAS, RGSB and HGKW).

The English Standard Version is clear when it says that the focus in Gen. 1:5 is on how God ordered time on a weekly cycle. This view is confirmed by comparing Scripture with Scripture— in this case, the meaning of *yôm* (day) in Genesis 1 with its usage and meaning in Ex. 20:8-11, and especially v. 11, where we read, "For in six days the Lord made the heavens and the earth, the sea, and all that is in them, but he rested on the seventh day."

Dr. John Morris, a Young Earth Creationist and president of the Institute for Creation Research, adds this further commentary:

> In this passage, God instructs *us* to work six days and rest one day because *He* worked six days and rested one day— during which week He created the heavens, the earth, the sea, and all things in them. The word "remember" in Hebrew, when used as a command, as it is in verse 8, refers back to a *real* historical event, and "for" in verse 11 is usually translated "because." It too refers back to a *real* historical event. Thus the days of our *real* work week are equated in duration to the *real* days of Creation. Same words, same modifier, same sentence, same slab of rock, same Finger which wrote them. If words mean anything, and if God can write clearly, then creation occurred in six solar days, just like our days.[3]

For the other exegetical and contextual arguments for the Church's long-standing belief in the literal twenty-four-hour meaning of "day" in Genesis 1 see page 151.

This is the first crucial issue that comes up in any discussion of Genesis 1 and 2. The meaning of "day" cannot be determined by scientific data or by forcing long ages into the clear meaning of Scripture. The meaning of "day," as we have shown, can only be determined by a careful internal examination of the scriptural passages in which *yôm* (singular) and *yamim* (plural) are used. The Bible is its own best interpreter.

The second crucial issue here is—

He Spoke and It Was Done

Because of their evolutionary bias or leanings, secular scientists and some Christian scientists, scholars and writers like Davis Young, Hugh Ross, and Francis Collins, believe that God, who is not bound by time and space, surely took more time than six literal days to create the universe, the solar system and man. Even Bible-believing Christians who love the Lord are so impressed with scientific opinions and theories that they are convinced that science must be compatible with Scripture, and so the two are combined somehow.[4] As it concerns evolution and old-earth ideas, this combination takes the form of Theistic Evolution, Progressive Creation, the Gap Theory, the Day-Age Theory, and the Intermittent Day Theory.

And yet, Genesis 1 and 2 speak unmistakably clear about God's creative activity in the beginning of time. In chapter one the absolute power of God is conveyed by the fact that He merely speaks and things are created. Each new section of the chapter is introduced by God speaking, "And God said, 'Let there be... '" followed by, "and it was so," or accomplished (1:3, 6, 9, 11, 14, 20, 24, and 26). Even the psalmist, centuries later, acknowledged the same thing when he wrote, "By the word of the Lord were the heavens made, their starry host by the breath of his mouth... For he spoke, and it came to be; he commanded, and it stood firm" (Psa. 33:6, 9).

Hebrew scholars point out that—

> In biblical thought God's word *(däbär)* and God's deed *(däbär)* are one in the same. Especially is this evident with respect to God. For His word is His creative word. To do was to speak, to speak was to do. God's word resulted in action.[5]

This evidence can only lead to one conclusion and that is that the creation account is told in terms of fiat (command) and immediate fulfillment.

The Symbiotic Evidence

Even though some scientists may agree with the biblical order of creation (light, plants, fish and fowl, and animals), they still contend that these created things appeared on earth over long periods (eons) of time, rather than in six successive literal days of twenty-four hours. Scott Huse argues,

> For example, if the days are actually ages, how did the fruit trees created on the third day survive for ages before the sun ["became visible at the surface of the earth"] on the fourth day? Similarly, this theory [the Day-Age Theory] fails to accommodate the vital symbiotic inter-relationships among plants (third day), birds (fifth day), and insects (sixth day).[6]

The writers of the *Hebrew-Greek Key Word Study Bible* argue in the same vein.

> Attempts to link the biblical account of creation with evolution are not supportable because the order of creation contradicts what would be scientifically necessary for an evolutionary process involving eons of time (e.g., the creation of plants and trees before the sun, which would

provide the light necessary for the photosynthesis that would cause vegetation to grow).[7]

The Literal Day Theory of Creation is the only position that logically supports and does justice to the biblical account of creation.

The Use of the Verb Bara *in 1:1, 1:21 and 1:27*

As a result of the mapping of the human genome, geneticist Francis Collins now claims to know the mind of God and understand how everything was created. In his book, *The Language of God,* he writes,

> Seeking to populate this otherwise sterile universe with living creatures, God chose the elegant mechanism of evolution to create microbes, plants, and animals of all sorts. Most remarkably, God intentionally chose the same mechanism to give rise to special creatures who would have intelligence, a knowledge of right and wrong, free will, and a desire to seek fellowship with Him.[8]

This statement is fraught with numerous biblical and theological difficulties and untruths. It is an arrogant assumption for a mere mortal (human) to claim to know the mind and will of the Almighty. How does he know that God deliberately chose the mechanism of evolution to create microbes, plants, animals, and man? What is so elegant about the mechanism of evolution when compared to the wondrous and mighty power of the eternal God? (Job 37:5, 14-24)

Collins' assertion is in direct contradiction to the exalted Hebrew prose, elegant style of the Mosaic cosmology of chapter one, and especially of the literary use of the Hebrew word *bara* in Genesis 1:1 (the creation of the universe), 1:21 (the initial creation of animal life), and 1:27 (the creation of man).

The verb *bara* denotes "that which is brought about (brought into existence) by the activity of divinity"— by God alone. Evolution had nothing to do with creation or God's creating activity. When God creates *(bara)* something, He does that which has no human parallel. His word or creative power effects His will, even creating the necessary material of which an object is to consist.[9] Finalizing this conclusion, we may say with absolute certainty, believing what the Scriptures reveal to us about God's essential attributes, especially His omnipotence and omniscience (all science), that God was not assisted in

His creation of the world and man by any natural forces, nor did He share His omnipotent power with naturalistic or evolutionary processes in order to create the world and man.

All Created Things were Mature at Creation

Ever since the theory of Darwinian Evolution broke on the world scene with the publication of his two books, *Origin of the Species* (1859) and *Descent of Man* (1871), atheists and naturalistic scientists have been captivated, if not obsessed, with this theory. They have even developed a hypothetical Geologic Column and evolutionary view of history which goes like this:

1. Most recent "Big Bang," 10-20 billion years ago.
2. Our solar system, 5 billion years ago.
3. Single-celled organisms, 3-4 billion years ago.
4. Multi-celled organisms, 1 billion years ago.
5. Humankind, 1-3 billion years ago.
6. Modern civilization, 5-10 thousand years ago.[10]

As John Morris states, "In education, in politics, in the media, and in social practice, evolutionary thinking has reigned supreme for the last several generations. It has even impacted church practice and the formulation of doctrinal understanding."[11] In fact, the evolutionary view of history is so pervasive and dominant today in American society and Christian circles that it has become, in the minds of many Christians, a permanent and acceptable *overlay* of Genesis 1 and 2. This is totally incongruent with divinely-revealed truth recorded in these two chapters. In other words, we now have a false worldview overlaying and strangling the true worldview given to Moses and to us. Because some Christians do not believe deeply enough in the inspired creation account in Genesis 1 and 2 and believing that science must have something to say about origins, they have been lured into a conflicted position and don't even realize it unless the subject comes up for discussion.

This conflict of worldviews is nowhere more apparent than in what the text really says and means concerning the things created. For example, Gen. 1:11 states, "Let the land produce vegetation: seed-bearing plants and trees… that bear fruit with seed in it, according to their various kinds." And it was so. Again in Gen. 1:20-22 we read, "So God created the great creatures of the sea... and every winged bird... according to its kind. And God saw that it was good. God blessed them and said, 'Be fruitful and increase in number and fill the water in the seas, and let the birds increase on the earth." The same thing is said of the land animals in Gen. 1:24-25 and of man and woman in Gen: 27-28; 2:7 and 21-23.

The only reasonable conclusion that we can come to is that everything that was created by the Word of God was created instantly mature at creation, capable of reproducing immediately, and that they were not created over long periods of time (millions and billions of years) by means of evolution. For a more complete list of aspects of a mature creation see page 83.

The Fixed Boundaries for the Reproduction of Plant and Animal Life

According to the general theory of evolution, the basic progression of life leading to man was in the following manner:

1. Non-living matter
2. Protozoans
3. Metazoan invertebrates
4. Vertebrate fishes
5. Amphibians
6. Reptiles
7. Birds
8. Fur-bearing quadrupeds
9. Apes
10. Man

If these things actually happened, it is perfectly logical and reasonable to expect that we should find vast numbers of transitional forms objectively preserved in the fossil record. This, however, is not the case and the supposed transitional forms between the major groups are missing in every case.[12]

Not only is there no evidence in Scripture that this evolutionary development ever took place, but we have the words of Charles Darwin who conceded in his writings that "Not one change of species into another is on record... we cannot prove that a single species has been changed."[13]

Vertical transformation (sometimes termed *macroevolution*) of one kind of organism into an entirely new organism is, however, prohibited and does not occur. Dogs never change into horses, Hawthorn plants never become roses, and finches never become anything but other finches.[14] Boundaries between kinds are very real and stubborn biological facts. When abnormal crosses are attempted, sterility is always the result. A horse crossed with a donkey produces a sterile mule, for example.

On the other hand, horizontal variation (sometimes referred to as *micro-evolution*), operating within limits specified by the DNA for the particular organism, is possible and has been used to develop many breeds. For example, there are over 200 varieties of dogs. It is possible to obtain 1500 varieties of the Hawthorn plant by natural variation. Darwin's various finches may also be attributed to such horizontal variation. Mankind even has considerable potential for variations into various races.[15]

Created organisms reproduce after their own kind and not after some other kind, with a limited amount of variation permitted within the permanently fixed kind just as Genesis 1:11, 12, 21, 24 and 25 records, and as God prescribed in the beginning of creation. This principle is also confirmed in the New Testament where we read in 1 Corinthians 15:38-39,

> But God gives it a body as he has determined, and to each kind of seed he gives its own body. All flesh is not the same: Men have one kind of flesh; animals have another, birds another and fish another.

Man, a Special Creation

For decades, since Darwin's idea of man's descent from a common ancestor became well-known in the late 1800s, anthropologists (having to do with man's origin), paleontologists (having to do with fossil remains), and evolutionists alike have been searching for the missing links between the apes and man, but these missing links have proven to be very elusive and non-existent. Yet, despite the lack of physical fossil evidence that man is descended from apes, the belief still persists that he is.

Some biologists now believe they have finally found impeccable evidence to prove this theoretical descent of man from apes. In 2003, under the leadership of Francis Collins, director of the Human Genome Project, the NIH's biomedical research team successfully finished mapping the complete sequence of several billion DNA subunits ("bases") and all of the genes that determine human heredity. This was followed in 2005 with the full mapping of the chimp genome by the same team which seemed to show a "near identity" with the human genome, with a ninety-five to ninety-nine percent match, depending on what factors are included.

This has led Dennis Venema, BioLogos senior fellow for science and the biology chairman at Trinity Western University, and other Christian scientists to conclude that "humans are not biologically independent, *de novo* creations, but share common ancestry" with prior primate species. Many biologists estimate that the biological branches separated from that common ancestor some 5 or 6 million years ago.[16]

Other Christian biologists, however, are "pushing back" on this scientific data and declaration. Biochemist Fazale Rana questions the ninety-five to ninety-nine percent figures, asserting that in any case common sense tells us "these types of genetic comparisons are meaningless" because they do not explain the "fundamental biological and behavioral differences" between chimps and humans. Rana also says close genetic similarity does not require shared ancestry.[17]

Needless to say, the above biological point of view linking man to chimps is in direct contradiction to the revealed truth in Genesis 1 and 2 that man is a special creation by God.

Genesis 2:7 clearly declares that "the Lord God formed the man from the dust of the ground and breathed into his nostrils the breath of life, and the man became a living being." Scientists of all stripes seem to be bent on discovering exactly how the invisible God could form a visible human being like you and me out of the dust of the ground. But is it possible to describe this marvelous happening in strictly biological and empirical terms? I think not. This supernatural happening belongs in the same category with the conception of the Son of God in Mary's womb, and the raising of the dead in the Old and New Testaments.

When it comes to spiritual and supernatural matters, God has not asked us as finite creatures to fully comprehend what He has done, but He does ask us to believe what He has done. "For my thoughts are not your thoughts, neither are your ways my ways, declares the Lord. As the heavens are higher than the earth, so are my ways higher than your ways and my thoughts than your thoughts" (Isa. 55:8-9).

Even though God has not chosen to reveal greater details about the creation of man, still He has revealed enough that we might believe that He did form the first human male. The first clue is found in Gen. 2:7 where we read the "the Lord God formed the man from the dust of the ground..." The verb "formed" (Heb. *yatsar)* conveys the picture of a potter's meticulous work in shaping a vessel from clay into a particular shape for a particular purpose (see Jer. 18:1-6). It is significant that the scripture (Gen. 1:24-25) does not say that God made the land animals this way. The second clue is found in the last part of verse 7, "...and [God] breathed into his nostrils the breath of life..." This is a recognition that life springs directly from God, whether it be in animals or in human beings. However, this life is totally different from the life animals received. Here God breathes life— physical, mental and spiritual— into the one created to bear His image (1:26-27). This is the beginning of man's spiritual nature.

The third clue that sets man apart from the animals is that man was made in the "image" and "likeness" of God (Gen. 1:26-27). This image is found chiefly in the fact that man is a personal, rational and moral being. While God is infinite and man is finite, nevertheless man possesses the elements of personality similar to those of the divine Person, thinking, feeling, and willing. Man was *created*, not evolved. This is expressly declared and this declaration is confirmed by Christ (Matt. 19:4). It is also confirmed by the unbridgeable chasm between man and beast. The highest beast has no God-consciousness or religious nature.

Finally, the Hebrew word translated "being" (verse 7) is *nepes,* which means "that which breathes." It corresponds to the Greek word *psychë* in the New Testament which is usually translated "life" or "soul." This verse argues against the idea that man evolved from some previously existing animal form and God merely breathed into that form the soul of man.[18] Thus, humanity is set apart from all other creatures (vv. 26-27) and crowned with glory and honor as the ruler of the earth (Psa. 8:5-8).

The last conclusion that grows out of our inductive and hermeneutical study of Genesis 1 and 2 is—

Creation is Completed

It is not uncommon today to hear astrophysicists and evolutionists claim that the universe is still expanding and that God is still creating. Theodosius Dobzhansky (1900-1975), a prominent scientist who subscribed to the Russian Orthodox faith and to Theistic Evolution has said, "Creation is not an event that happened in 4004 BC; it is a process that began some 10 billion years ago and is still underway..." [19]

Yet the Scripture declares, "Thus the heavens and the earth were completed in all their vast array." How comprehensive is that? It sounds limitless! "By the seventh day God had finished the work he had been doing; so on the seventh day he rested from all his work" (Gen. 2:1-2, see also Ex. 20:11; 31:17; Heb. 4:3-4). One commentator writes, "The grand design of the heavens and the earth was brought to perfect completion. The host, or more likely, 'the army' of the heavens and the earth was in its place."[20]

From this brief explanation of Gen. 2:1-2 it seems reasonable to believe that God's activity of creating the heavens (and very possibly the whole universe) and the earth was fully completed in six days. Beyond that one can only speculate.

The Conclusions Summarized

1. The only meaning for day (Hebrew *yôm)* in Genesis 1 is that of an ordinary day of twenty-four hours.
2. God created everything on earth by fiat, or command.
3. The symbiotic interrelationships among plants, birds, and insects require a twenty-four-hour day.
4. The use of the Hebrew word *bara* in Gen. 1:1; 1:21 and 1:27 militates against evolution and upholds divine Creatorship alone.

5. Everything brought into existence by the spoken word of God was mature at creation.

6. There are fixed boundaries for the reproduction of plant and animal life that cannot be crossed just as the Bible claims, "according to its kind."

7. Man is a special creation of God, completely different from all the other creatures. This is so because man bears the image of His Creator, which is found in the fact that man is a personal, rational and moral being. Like his Maker, man possesses the elements of personality such as thinking, feeling and willing. This spiritual nature and image of God was infused into Adam when God breathed life into him, and this elevated nature that sets man apart from the beasts was passed on to Adam's posterity through the hereditary process.

8. From the plain meaning of Gen. 2:1-2, it seems that creation is definitively finished.

These cumulative conclusions are a strong argument for a young earth.

Endnotes

1. Thom Rainer, "And the Lord Added to Their Number Daily... Evangelism and Post-modern Culture" in *Southern Seminary*, Spring 1997, Vol. 65, No. 2, 12.

2. Loc. cit.

3. John D. Morris, *The Young Earth* (Green Forest, AR: Master Books, 1997), 29.

4. Ibid, 34.

5. Eugene Carpenter, "Cosmology" in *A Contemporary Wesleyan Theology*, vol. 1 (Salem, OH: Schmul Publishing Co., Inc., 1992), 158.

6. Scott M. Huse, *The Collapse of Evolution* (Grand Rapids: Baker Book House, 1988), 32.

7. Spiros Zodhiates, ed., *Hebrew-Greek Key Word Study Bible* (Chattanooga, TN: AMG Publishers, 1996), 2.

8. Francis S. Collins, *The Language of God* (New York: Free Press, 2006), 201.

9. Carpenter, op. cit., 147-148.

10. Morris, op. cit., 38.

11. Foreword to my book *Creation for Earnest Believers* (Nicholasville, KY: Schmul Publishing Company, 2012), 9.

12. Huse, op. cit., 43-44.

13. Quoted in *The Collapse of Evolution*, 38.

14. Huse, op. cit., 40.

15. Ibid, 39.

16. Richard N. Ostling, "The Search for the Historical Adam," in *Christianity Today*, June 2011, 25.

17. Loc. cit.

18. Zodhiates, op, cit., 4.

19. Collins, op. cit., 206.

20. Lee Haines, "Genesis" in *The Wesleyan Bible Commentary*, Vol. 1 (Grand Rapids: William B. Eerdmans Publishing Company, 1967), 30.

12

Wanted: A Centurion-like Faith – "Just Say the Word"

Introduction

IN THIS CLOSING CHAPTER, I want to leave the reader with a remarkable, powerful and reasoned essay by John Hultink about the unusual encounter Jesus, the Son of God, had with the Roman centurion and the instantaneous healing of his servant who "was sick and about to die." Many thanks to Joel Belz and WORLD for this brief introduction of John.

Into the continuing discussion of the earth's origins in general and theistic evolution in particular comes John Hultink, a native of the Netherlands who has lived most of his life in St. Catherines, Ontario, Canada.

A WORLD subscriber and backer since its first issue, John is a real estate developer, book distributor, and publisher of books and newspapers. He has a keen interest in Christian education, philosophy, theology, and the application of biblical truth to all of life.

John graciously has allowed WORLD to publish his thoughts on how creation bears testimony against theistic evolution.— Joel Belz"

The Essay[1]

Jesus is not given to exaggeration. Yet what he said about the Roman centurion recorded in Matthew (8:5-13) and Luke (7:1-10)... appears to lean in that direction. Jesus said this centurion had faith that was greater than the faith of anyone in Israel. This claim would entail that the faith of this non-Israelite was greater than the faith of Peter, James, John, and the believing members of the synagogue. Such faith would bring this man into the company of Old Testament believers like Job and Abraham. Yet this centurion did not know Jesus personally. He was not even an Israelite. He did not consider himself worthy even to ask Jesus to come into his house. So what made his faith so remarkable?

This man, this Gentile believer no less, had faith that was so exceptional that both Matthew and Luke recorded that Jesus marveled at this faith. Yes, marveled. That is, Jesus met tens of thousands of people during His ministry but seldom did He have occasion to marvel at someone's faith.

What was so exceptional in this instance? What was there about this man's faith for Jesus to marvel at? A great deal! And Jesus used this incident to teach His disciples, Israel, and us an important lesson. Even today, 2000 years after the event, Jesus uses his encounter with the Roman centurion to drive home the lesson of what it means when Scripture emphasized that He, Jesus Christ, is "the Word." In an age where the knowledge derived from the field of science has been elevated to idolatrous heights, we have lost awareness and sensitivity to the unique nature and power of "the Word."

This centurion believed that Jesus also is a Man who exercised authority, authority infinitely greater than his own. He believed Jesus has the awesome authority to command all aspects of creation. All Jesus needs to do to exercise this authority according to the centurion is to say the Word, and his servant would be healed. No questions, no calculations, no doubts. What a tremendous faith. As children place their trust in their father and mother, so this Gentile believer placed absolute trust in Jesus. Just say the word, Lord, and sickness, disease, and death will flee at your command, he fervently believed. All Jesus has to do is speak the command. This Roman centurion would have had no difficulty believing the biblical record that Jesus spoke the creation into being by the power of His Word.

Yes, the centurion believed that Jesus' authority is that great: Just say the Word, Lord, and my servant will be healed. And that, said Jesus, is an expression of faith that should fill the hearts of all believers: But I have not found faith like that anywhere, no, not even in all Israel among the hundreds and thousands of descendants of Abraham, Isaac, and Jacob (Mark 6:6).

In the depth and breadth of his faith, this centurion acknowledged Jesus as Creator and Lord over all creation. God enabled this man to understand the "servant character" of creation. He acknowledged that Jesus is creation's Lord and has the power to command the entire creation to do His will. Even as the centurion's underlings obeyed when he said, "Come" and "Go," so also creation instantly obeys the command of its Sovereign and Lord. No questions asked, not a moment's hesitation, no backtalk: Just say the

Word, Lord. Just say the Word, and your will is accomplished. That, said Jesus, is faith upon which His kingdom is built. If Christian professors and students at universities and seminaries around the globe would believe Scripture with the same unquestioning faith, if they would lead lives of uncompromising trust and faith and view the origin and unfolding of creation with the unshakable conviction that God in Jesus Christ is absolute Sovereign over creation, well, then, the pulpits would reverberate with the Word of power and the people of God would rejoice till the rafters shook. Even unbelievers would marvel at such faith.

God indeed commands His creation as His servant. The servant character of creation is foundational to the Christian's faith— even as the centurion believed, just think what that means for our understanding of the creation account in Genesis. The same Jesus who commanded the centurion's servant to be healed was present as God "in the beginning." And in the beginning He commanded the creation into existence exactly as described in the various books of the Bible. That is, by the power of Christ's "Let there be." A description that must be accepted in faith because the act of creation surpasses our understanding. No one can search out the mechanics of the nature and power of Christ's spoken Word. Why? Because the "mechanism" of creation is the Word. The power that brought the creation into being on command is the same power that had "gone out" of Jesus and gave life to the touch of the woman with the debilitating issue of blood (Matthew 9:18-22; Mark 5:21-34). The power that transformed nothing into a universe of millions of objects constructed out of an array of atoms and elements is the same power that commanded the stinking corpse of Lazarus to rise and walk out of the grave: a re-created Lazarus identical in appearance to the predeceased Lazarus and with the complete memory of a lifetime restored (John 11:38-44).

The spoken Word in action. To repeat: No one can search out the mechanics of the originating, creating activity of the spoken Word. There is no possibility of reverse engineering here. The finished creation, which is a manifestation of the Word objectified, is the proper object of human inquiry— the act of creation is not. The Word, after it comes to visible expression as creation, is man's proper field of study, not the creative activity of the Word itself. And once we try to understand the finished creation, that is, try to understand what is referred to in some circles as the "creation order," we should do so with the acknowledgement that this order of creation came into being solely through the creation acts of God. And God sovereignly determines the order of His creation acts. Once again, the creation acts of God cannot be subjected to human analysis. To attempt to do so would be to attempt to subject God to our finitude. Here caution must be exercised, for that is blasphemy.

Genesis 1 is God's revelation of his almighty creation acts: "Let there be." "Let there be." "Let there be." Genesis 1 as the revealed record of God's creation acts is not ours to analyze any more than we can analyze the resurrection of Lazarus. What physician of sound mind would do a post-resurrection examination on Lazarus to determine what brought him back to life and how? So also with the 10 "Let there be" commands of Genesis 1.

Christians who stumble over the powerful and magisterial "Let there be" of Genesis 1 will stumble again when they meet Christ in the New Testament. In the New Testament they hear Christ commanding, "Let there be," again and again. This time, if they persist in their unbelief, they will not rise again. The New Testament will become a snare to them, a how-to book of morals and personal conduct, a book of health and wealth. The sovereignty of God and the creative, unfolding, upholding power of Jesus' "Let there be" will remain a mystery. The New Testament will then be a closed book that cannot possibly end in a mighty, universal resurrection when all the inhabitants of the earth respond to Christ's command: "Lazarus, come forth."

Contemporary Christians under the blinding deception of the pretended autonomy of science have fallen into the destructive habit of reading scripture as a collection of metaphors. We no longer know how to take God's Word seriously. We have lost sight of the power of the Word. We have in important instances needlessly metaphorized the Word. We have lost sight of who Jesus is and who Adam could have been. We read in Job 9:8 that God "treads on the waves of the sea," and we are inclined to say, "Oh, that is merely Scripture's way of poetically portraying God's power. We are not meant to take the poetic genre literally; God never actually deigns to walk on the waves of the sea, not literally. Does He?"

Scripture tells another story. At the height of a storm Jesus' disciples, seasoned fishermen feared they would perish in a raging sea. But Jesus rebuked the winds and the sea and commanded a great calm (Matthew 8:23-27; Mark 4:35-41; Luke 8:22-25). That is, the wind and the sea listened and obeyed their Master. They would have obeyed a faithful first Adam as well. On another occasion (Matthew 14:22-33, Mark 6:45-52; John 6:16-21), when the disciples were in their boat in the middle of a storm, Jesus, who had gone into the mountains to talk with His Father, approached them walking on the sea. If Peter had exercised faith in measure to the centurion's, he also would have been able to walk on the sea. It was his unbelief that sank him, not his lack of authority over the creation.

Metaphors apply to humans but seldom to God, other than to help make God's acts in history comprehensible. Metaphors often engage in overstatement, but it is not possible to overstate God's abilities. God's power over His creation is awesome. Job spoke prophetically when he declared, "God trampled the waves of the sea." When the Son of God clothed Himself with flesh, He did just that. He walked on the waves of the sea for His disciples to see and to remember and to witness to you and to me. Failure to comprehend the servant character of creation skews our understanding of the entire Bible and minimizes our ability to glorify God. There are no metaphors in play here on the boisterous sea. What we see is what Abraham heard, "I am God Almighty" (Genesis 35:11).

Do not fall into Satan's trap of unnecessarily metaphorizing Jesus' Word into a powerless literary device. Scripture may not actually be speaking metaphorically even if we think so. The Word of God is the Person of the Trinity through whom the entire creation was spoken into being. And that Word continues to uphold the creation to this day. Think of "summer and winter and springtime and harvest." Think of the "sun, moon, and stars in their courses above." Christ can credibly declare, if such a declaration were necessary, that the "mechanism" that brought the creation into being is His spoken Word: "Let there be." But that mechanism, that creative Word, is not subject to human analysis. No one can express in scientific terms (or non-scientific) how creation came into being. What is subject to human analysis is the completed structure of creation as it confronts us, not God's act of "creating" itself. For that reason alone the source of life will never be discovered by science (knowledge discovered) because all life originates with God and God alone (John 5:24-29; Genesis 2:7).

And it makes no difference whether Christ, in the act of creation, is creating out of nothing or out of preexisting material. (Material some commentators believe they find in Genesis 1:2). In the case of the re-creation of Lazarus, Christ was working with maggot-infested, rotting material. It made no difference compared to creation out of nothing in the beginning. The mechanism (or process) that brought Lazarus back to life instantaneously was the creating power of the Word: "Let there be." "Lazarus, come forth."

We, the living, can today experience and witness the creative power of the Word. As a wolf-life creature "evolves" (evolution within the species) into 600 North American species of dog before our eyes, so it is that the creative Word of God continues to unfold before our eyes.

Seven billion people today inhabit the earth, people of different nations, colors, physical appearances. Few of these 7 billion people look alike, as one would expect if they were the product of some mindless cookie-cutter process of evolution. What we actually witness in the world around us is the dynamic potential God placed in creation in the beginning coming to actualization before our eyes.

Even as the spoken Word healed the centurion's servant instantly and raised Lazarus from among the dead, in the blinking of an eye, so the Word is revealed to operate in power throughout Scripture again and again and again. And so it operated with power in the beginning. And again when Christ became One of us and lived among us. It is Christ's exercise of this awesome, creative power that reveals the character of God. And for those who have eyes to see, this power of God is still manifested today, for instance, as one kernel of Iowa seed corn is planted in the earth and in a few months' time bursts into a cob bearing 800 kernels in 16 rows. Where did that creative power with the seed corn come from?

"Let there be," God said in Christ in the beginning and again 2000 years ago in the Person of Jesus Christ. Literally. For there is no other way for Christians to understand the creative power of God, even as creation now continues to open itself up to us, but to accept in centurion-like faith that God said, "Let there be." And behold: It was.

God does not owe theistic evolutionists an explanation. Their contention that God is the author of a messy process of evolution birthed by mysterious physic-chemical chance processes resulting in the malfunctioning or malformation of existing perfect genes and driven by a natural selection of whatever it is that is out there to get selected over a period of billions of years borders on blasphemy. God thereby becomes the servant of so-called autonomous processes. It makes God the subordinate author of design by death and failure and chance and time.

Lazarus and the son of the widow from Nain as well as the daughter of Jairus, all resurrected on command, testify against this monstrous falsehood. God is not party to a wretched neo-Darwinian process of evolution. As sovereign Creator, His commanding Word, and His Word alone, is the mechanism that brought the universe into being. The entire creation bears testimony against theistic evolutionism, against an evolving that is no more than a parasite that feeds on God's creation.

For God, even to this day, continues to speak to us clearly through His Word and deeds, behold the riveting beauty of creation each year anew as God's nature bursts into spring, as eggs hatch and the newly born eagle spreads its majestic wings. Life, a burst of growth and exotic color everywhere. My God, how great thou art! Behold and tremble in awe all you peoples of the earth, for God did a wonder-filled thing when He spoke a pregnant (after their kind) creation into being. If only God would give us the eyes of the centurion.

Endnote

1. *http://www.worldmagcom/2013/08/just_say_theword/page4.* Used by permission.

Afterword

"There is a way that seems right to a man, but in the end it leads to death" —Proverbs 14:12

THE BIBLE FROM BEGINNING to end sets two ways before each of us:
- The way of life and the way of death (Deut. 30:19-20).
- God's way and our own way (Isa. 48:17 and 53:6).
- The way of the righteous and the way of the ungodly (Psa. 1).
- There are only two worldviews— the one that is God-centered and the one that is man-centered.

These worldviews are based on divine *revelation* and human *reason* respectively.

It is interesting that these are polarizing views— that is, they are poles apart. It is also worthy of note and careful consideration that there is no "happy median" or "mediating position" between these two ways— none whatsoever. It is either life or death, light or darkness. In life, God has given each one of us freedom of will to choose between these two ways and what we will believe. In the end, He will hold us accountable for the way we have chosen and what we have chosen to believe (2 Cor. 5:10 and Prov. 23:7 KJV).

In the body of this book I have explained Charles Darwin's theory of biological evolution from non-living matter to the final realization of man. It is gratifying to know that less than ten percent of the American public believes in the official scientific orthodoxy which is that humans (and other living things) were created by a materialistic evolutionary process in which God played no part.

I also set before the reader the opposing view which is timeless. It is the conservative biblical view of Genesis 1 and 2 that man came from the hands of the Lord God (Gen. 2:7) and that he was made in the image of God his Maker (Gen. 1:26-27). This is the unmistakable teaching of the Genesis narrative and other portions of Scripture.

Polls show that forty-five percent of Americans accept this belief as Biblical Creationists.

Then I spent considerable time outlining the current Theistic Evolutionary theory of origins, which is a so-called "mediating position" between Darwinian orthodoxy and Biblical Creationism. It is the belief in God as Creator and that he used evolution as His method of creating. This view is held by the other forty-five percent of Americans. The reason why this view seems so popular to so many, and increasingly attractive even to evangelical intellectuals, is that is seems like a "perfect solution" to the potential conflicts between twenty-first-century science and religion, meaning the Christian faith rooted in the revealed creation facts of Genesis 1 and 2. Theistic Evolution seems to provide that "bridge" that links the spiritual world with the scientific world. In other words, it's the best of both worlds. Or so it seems. But not everything in life that seems right and good is as it seems.

Satan's proposal to Eve was that she could eat of the tree of the knowledge of good and evil and not die. She could even be like God. She forgot that she was already a finite version of God. So it seemed all right for her to eat of the prohibited fruit. She could have her cake and eat it too. Or so she thought. But we know how that turned out. The Bible very clearly states, "There is a way that seems right to a man, but in the end it leads to death" (Prov. 14:12).

Despite Samuel's warning, Saul, the first king of Israel, thought that he could spare King Agag and the best of his flocks and still remain king. That seemed all right for him to do without suffering any consequences. Or so it seemed. But because of his disobedience (rebellion) the kingdom was torn from him and he ended up taking his own life.

Then there is the story of David who thought he could have a "fling" (affair) with Bathsheba and nobody would ever know about it. So he covered it up for a whole year. Finally, he confessed his double sin to God and was restored to fellowship with Jehovah. But King David paid a high price for his transgression. He and Bathsheba not only lost their child, but things were never the same again in David's family.

So, in the light of divine truth mentioned above— namely that there are only two ways in life, two worldviews, one based on revelation and the Bible and the other on man's darkened reason, and that there is no mediating position between them— let us examine the concept of Theistic Evolution very carefully to see if it is true or specious (having deceptive attraction or allure), and if it is genuine or has a false look of truth. In this evaluation, we will use the Genesis 1 and 2 account of origins and that they occurred some 6,000 to 12,000 years ago, as our "north star" or point of reference in judging the validity of Theistic Evolution.

Illusions of Theistic Evolution

First of all, philosophically and theologically, Theistic Evolutionists are laboring under a number of illusions with regard to their theory.

The *first illusion* is that Theistic Evolution as a "mediating position" between Creationism and materialistic evolution exists, or that it can be supported by empirical evidence. The opening words of Genesis, "In the beginning God created the heavens and the earth," are a statement given by divine revelation and accepted by human faith. It is beyond the province of empirical science to either prove or disprove them. No equivalent theory of ultimate origins has ever been offered. It should be noted that the Genesis account of origins has existed since the beginning of time. The theory of Theistic Evolution only arose in the nineteenth century as an offspring of Darwin's theory of biological evolution. Furthermore, there is absolutely no empirical evidence that God chose the mechanism of evolution to create all things as Francis Collins and other Theistic Evolutionists claim.

The *second illusion* is that there is a need at all to harmonize science and the Scripture. Who is calling for this harmonization? It is the scientists and the Theistic Evolutionists, who because of their beliefs and biases, feel that they have something to contribute to the revealed story of origins— not Creationists and most people of faith. In 1954 Bernard Ramm was adamant about the imperative necessity for a harmony of science with Scripture. More recently, Francis Collins went beyond this in his book to suggest that men of faith and science explore a pathway toward a sober and intellectually honest integration of these views (the scientific and the spiritual) even as he admits that these are two separate domains.

What is there to harmonize? God has already given us (humanity) the only record of origins and creation that exists. Science is a newcomer. Furthermore, although Genesis does not purport to be a textbook of science, nevertheless, when it touches upon scientific subjects, it is accurate. Science has never discovered any facts which are in conflict with the statements of Genesis 1.

The *third illusion* is that it is possible to explore a pathway toward a sober and intellectually honest integration, or synthesis, of these two views— the scientific and the spiritual. If these are two separate domains, how can they be integrated honestly? Such integration is unrealistic for, as Dr. Albert Mohler correctly states, "in the process of attempting to negotiate this new middle ground [Theistic Evolution], it is the Bible and the entirety of Christian theology that gives way, not evolutionary theory."

The *last illusion* is that Theistic Evolutionists and some Christian scientists believe that they can arrive at a knowledge of the origins of man and the universe by scientific discovery alone, independently of exercising faith in that which is already known and created, namely God's wonderful handiwork on display in our world. As someone has pointed out, both evolution and Christianity are religions that cannot be proved scientifically (empirically) and both must be accepted by faith. Faith in the Christian doctrines is the kind of faith that is spoken of in Hebrews chapter eleven where we read,

> Now faith is being sure of what we hope for and certain of what we do not see. This is what the ancients were commended for. By faith we understand that the universe was formed at God's command so that what is seen was not made out of what was visible. By faith Abel offered God a better sacrifice than Cain did. By faith he was commended as a righteous man, when God spoke well of his offerings. And by faith he still speaks, even though he is dead. By faith Enoch was taken from this life, so that he did not experience death; he could not be found, because God had taken him away. For before he was taken, he was commended as one who pleased God. And without faith it is impossible to please God, because anyone who comes to him must believe that he exists and that he rewards those who earnestly seek him." (Heb. 11:1-6)

My question to the reader is simply this: Can God commend us today in this scientific age for this kind of faith— of being sure of what we hope for and certain of what we do not see with our natural eyes under a microscope?

The ancients (people of old), like us, were not present when the universe and man were created at God's command, yet they were certain of that which they were not witnesses to. By faith they understood that they— man and the world— came into existence by God's will, power and command, for they had received this revealed truth by oral transmission from previous generations. They never doubted for a moment that it had happened or that it had happened in this way.

Appendix A

A Model Creation Statement from Appalachian Bible College[1]

W E BELIEVE that the first eleven chapters of Genesis are the literal history of the early Earth (Matthew 19:4, 24:37). We believe that this material universe is the result of a sequence of unique creative acts of God the Son, accomplished with the aid of God the Holy Spirit and directed by God the Father (Genesis 1:1, 2; Colossians 1:16). We believe these creative acts were *ex nihilo*, completed by the mere spoken commands of God (2 Peter 3:5). We further believe that these creative acts were accomplished in six literal twenty-four hour days (Exodus 20:11). Therefore we hold to a young earth view supported by the genealogies and other time information provided in the Word of God. We also believe that the material universe was created in total perfection (Genesis 1:31) but subsequently was sentenced to a slow decay and eventual destruction by the Curse (binding), which was part of the penalty for the disobedience of the parents of all mankind, Adam and Eve, whom we view as real, literal people, created on the sixth day of Creation (Genesis 1:27, 2:7-3:19). We reject all concepts of a pre-Adamic race. We believe that the biblical Noahic Flood (Genesis 6-8) was a real, year-long global event, the result of the judgment of God on the hopelessly rebellious descendants of Adam and Eve (Genesis 6:5, 1 Peter 3:6), and resulted in much of the present geology of the Earth, including most of the fossil graveyards of myriads of plants and animals then living. We believe that only eight human souls, Noah and his family, survived the Flood (Genesis 7:13 and 8:18) and that all mankind now living are descended from this family, dispersed over the face of the Earth by the confusion of tongues described in Genesis 11.[2]

Endnote

1. Located in Beckley, West Virginia, in the beautiful Appalachian mountains.

2. Used by permission and quoted in *Already Compromised* by Ken Ham and Greg Hall, 31.

Select Bibliography

Books

Beer, Gavin de. *Charles Darwin*. London: Nelson and Sons Limited, 1963

Bromall, Nick. *Biblical Criticism*. Grand Rapids: Zondervan Publishing House, 1957.

Cairns, Earle E. *Christianity through the Centuries: A History of the Christian Church*. Grand Rapids: Zondervan Publishing House, 1958.

Cannon, William R. *The Theology of John Wesley*. Nashville: Abingdon-Cokesbury Press, 1946.

Collins, Francis S. *The Language of God*. New York: Free Press, 2007.

Darwin, Charles. *The Autobiography of Charles Darwin, 1809-1882*. New York: W. W. Horton and Co., 1958.

Dowley, Tom, ed. *A Lion Handbook: The History of Christianity*. Herts, England: Lion Publishing, 1977.

Ferré, Frederick. *Language, Logic and God*. New York: Harper and Row, 1961.

Forté, Marius and Sam Sorbo. *The Answer: Proof of God in Heaven*. Telemachus Press, L.L.C., 2013.

Ham, Ken and Greg Hall. *Already Compromised*. Green Forest, AR: Master Books, 2011.

Henrichson, Walter A. *Principios de Interpretação da Biblia*. São Paulo: Editora Mundo Cristão, 1997.

Henry, Carl F. H. *Remaking the Modern Mind*. Grand Rapids: Wm. B. Eerdmans Publishing Company, 1946.

Huse, Scott R. *The Collapse of Evolution*. Grand Rapids: Baker Book House, 1988.

Johnson, Phillip. *Defeating Darwinism by Opening Minds*. Downers Grove, IL: InterVarsity Press, 1997.

Kaufmann, Donald A. *Eleven Tablets and a Papyrus Scroll: A Creation Commentary of Genesis Tablets 1-6, vol. 1*. Kissimmee, FL: Ktizo Publications, 2008.

King, C., Wesley. *Creation for Earnest Believers*. Nicholasville, KY: Schmul Publishing Co., 2012.

Livermore, Paul. *The God of Our Salvation, 2 vol.* Indianapolis, IN: Light and Life Press, 1995.

Livingstone, Elizabeth A. ed. *The Concise Dictionary of the Christian Church*. Oxford: Oxford University Press, 1987.

Matrisciana, Caryl and Roger Oakland. *The Evolution Conspiracy*. Eugene, OR: Harvest House Publishers, 1991.

Morris, Henry. *Studies in the Bible and Science*. Philadelphia: Presbyterian and Reformed Publishing Co., 1967.

Morris, John D. *The Young Earth*. Green Forest, AR: Master Books, 1997.

Muncaster, Ralph O. *The Bible: Scientific Insights*. Mission Viego, CA: Strong Basis to Believe, 1996.

Nash, Ronald H. *The New Evangelicalism*. Grand Rapids: Zondervan Publishing House, 1963.

Ramm, Bernard. *The Christian View of Science and Scripture*. Grand Rapids: Wm. B. Eerdmans Publishing Company, 1955.

Schaffer, Francis. *The Great Evangelical Disaster*. Westchester, IL: Crossway Books, 1984.

Strobel, Lee. *The Case for a Creator*. Grand Rapids: Zondervan, 2004.

Taylor, Ian T. *In the Minds of Men: Darwin and the New World Order*. Toronto: TYE Publishing, 1984.

Traina, Robert A. *Methodical Bible Study*. No Publisher, 1968.

Vos, Howard. *Exploring Church History*. Nashville: Thomas Nelson Publishers, 1994.

Young, Edward J. *An Introduction to the Old Testament*. Grand Rapids: Wm. B. Eerdmans Publishing Co., 1950.

Articles

Brown, A. Philip. "Origins Debate." *God's Revivalist and Bible Advocate* (May 2014).

Brown, Collin. "The Ascent of Man." *A Lion Handbook: The History of Christianity*. Herts, England: Lion Publishing, 1977.

Carpenter, Eugene. "Cosmology" in *A Contemporary Wesleyan Theology*. Salem, OH: Schmul Publishing Co., Inc., 1992.

Coppenger, Mark T. "Evolution and Creation in Higher Education." *Southern Seminary* (Winter 2011, vol. 79, No. 1).

Dailey, Jim. "Dismantling Darwinism: A Conversation with Phillip E. Johnson." *Decision* (August 2003).

Hammond, Al. "A Missionary Speaks Out on Science/Faith Conflicts." *Mission Frontiers* (January-February 2004).

Honey, Charles. "Adamant on Adam." *Christianity Today* (June 2010).

Mohler, R. Albert Jr. "A Letter from the President." *Southern Seminary* (Fall 2010 Vol. 78, No. 4).

—————. "The New Atheism and the Dogma of Darwinism." *Southern Seminary* (Winter 2011, Vol. 79, No. 1).

―――――. "The Inerrancy of Scripture: The Fifty Years War... and Counting." *Southern Seminary* (Fall 2010, Vol. 78, No. 4).

―――――. "The New Shape of the Debate." *Southern Seminary* (Winter 2011, Vol. 79, No. 1).

Oaks, Stan. "Toppling the Giant" *Campus Alert*. Carrollton, TX: Christian Leadership Ministries, a ministry of Campus Crusade for Christ. (Vol. 5, No. 2 January 1997).

Ostling, Richard N. "The Search for the Historical Adam." *Christianity Today* (June 2011).

Price, G. M. "Revelation and Evolution: Can They Be Harmonized?" *Journal of the Transactions of the Victoria Institutes, 1925.*

Reeves, Carolyn, "When they can't see the Forest for the Tree" *AFA Journal* (July 2009).

Schofield, R. E. "The Lunar Society of Birmingham." *Scientific American 247*, (June 1982).

Siefken, Hugh. "Faith in the Physics Lab." *Light and Life* (August 1996).

Stafford, Tim. "A Tale of Two Scientists." *Christianity Today*. (July/August 2012).

Van Biema, David. "God vs. Science: A Spirited Debate between Atheist Biologist Richard Dawkins and Christian Geneticist Francis Collins". *Time* (November 13, 2006).

Wills, Gregory A. "Creation and American Christianity." *Southern Seminary* (Winter 2011, Vol. 79, No. 1).

Reference Works

Alexander, David and Pat Alexander, eds. *Eerdmans' Handbook to the Bible*. Grand Rapids: William B. Eerdmans Publishing Company, 1976.

Barker, Kenneth, ed. *Reflecting God Study Bible*. Grand Rapids: Zondervan Publishing House, 2000.

Brown, Francis, S. R. Driver and Charles A. Briggs. *Hebrew and English Lexicon of the Old Testament, 9th printing*. Peabody, MA: Hendrickson Publishers, 1906.

Carter, Charles, ed. *A Contemporary Wesleyan Theology*. Salem, OH: Schmul Publishing Co., Inc., 1992.

Carter, Charles, ed. *The Wesleyan Bible Commentary, Vol. I*. Grand Rapids: William B. Eerdmans Publishing Company, 1967.

ESV Study Bible, English Standard Version. Wheaton, IL: Crossway, 2008.

Ferm, Vergilius, ed. *An Encylopedia of Religion*. New York: The Philosophical Library, Inc., 1945.

Gingrinch, F. Wilbur and Frederick W. Dunken. *Lexico do N.T. Grego-Portugues*. São Paulo: Sociedade Religiosa Edições Vida Nova, 1984.

Harper, Albert, ed. *The Wesley Bible: A Personal Study Bible for Holy Living*.

Nashville: Nelson Publishers, 1990.

Koehler, Ludwig and Walter Baumgartner. *Hebrew and Aramaic Lexicon of the Old Testament*, Volume 1. Leiden; Boston, MA: Brill, 2001.

Life Application Study Bible, NLT. Wheaton, IL: Tyndale House Publishers, Inc., 1996.

Metzger, Bruce M. *Lexical Aids for Students of New Testament Greek.* Princeton, NJ: Published by the author.

Ryrie, Charles Caldwell. *The Ryrie Study Bible, NAS.* Chicago: Moody Press, 1978.

Scofield, C. I., ed. *Holy Bible: The New Scofield Reference Bible.* New York: Oxford University Press, 1967.

The Hymnal for Worship and Celebration. Irving, TX: Word Music, 1989.

Webster's Ninth New Collegiate Dictionary. Springfield, MA: Merriam Webster, Inc., Publishers, 1956.

Zodhiates, Spiros, ed., *Hebrew-Greek Key Word Study Bible.* Chattanooga, TN: AMG Publishers, 1996.